The Lunar Effect

Harry Alcock

MOANA PRESS

Cover illustration courtesy of
U.S. Information Service.
Photo taken from Command Module *Columbia*.

ISBN 0-908705-63-8
Published by Moana Press
19 Roderick Street
Tauranga, NEW ZEALAND
Phone (075) 63-011
and 112 Pope Street
Plimmerton, Wellington
NEW ZEALAND
Phone (04) 331-842

Printed and typeset by
Printcorp Services Ltd, Tauranga

CONTENTS

Contents (continued)

To my wife, Dulcie.

For more than 20 years Dulcie has listened to me theorising on the Moon's influence on the weather. She has typed hundreds of thousands of words over the years on a subject not totally understood by her.

Then, to make life even more difficult, Dulcie had to master the intricacies of a word processor at the end of a working life when many would not have bothered to do so.

But I am sure that she appreciated its assistance in the long term.

Also, my thanks to the *Waikato Times* for its support over the years.

Countries throughout the world will have spent thousands of millions of dollars in an attempt to understand the atmosphere and the way it behaves.

In the earlier days of meteorology, manned balloon flights were undertaken with the pilots and scientists facing extreme dangers at the high altitudes the balloons attained in an effort to find out what happened up there.

In more recent times, rockets were sent up to altitudes not attainable by way of the balloon flights and extra knowledge was gained on the behaviour of the atmosphere.

From the knowledge obtained, meteorologists devised model atmospheres in an attempt to explain its movement. This was followed by complicated equations which further attempted to explain the movements of the atmosphere.

Meteorologists have always dreamed that one day they would be able to offer a long range weather forecast and a form of numerical forecasting was indeed developed, but without a great deal of success.

Today the emphasis is on computer analysis of all the factors involved. But, after about 150 years, meteorologists still have not completely unravelled the mysteries of the atmosphere and its associated weather. Progress certainly has been made, but without any true understanding on what makes the atmosphere circulate in the manner that it does.

This book offers another theory, one which the majority of meteorologists and climatologists have chosen to ignore or perhaps it may be more truthful to say, they did not know existed.

Some analysis has been done on the Moon's influence on our weather and the results may be considered inconclusive by many and by others completely negative.

The theories propounded here may penetrate some small cracks in a dam and widen them to allow this new knowledge to flood forth on the true method which explains how the atmosphere circulates as it does, and how it affects our weather.

It is not expected that everyone will accept this theory completely, some may be prepared to go along with it in part and then wait for a more complete analysis or detailed research before a possible reserved agreement with the principles stated.

1: *Meteorology*

Where is it heading?

The greatest advance in the sciences in the history of man has been in the 20th century, with the exception possibly of one — the science of meteorology.

Meteorology has developed slowly after it was discovered by a Frenchman in 1648 that air had pressure. After this discovery and research into the earliest mercury barometers, how could they be used beneficially in meteorology? It took some time for the this new instrument to be recognised for what it was, i.e. an instrument for predicting weather changes.

With continuing research meteorology became established more or less in its present form earlier in this century, but except for slight changes in expression — such as the term 'west type depressions' to 'trough of low pressure' and the renaming of the high latitude cyclones to depressions — and incorporating modern aids such as satellite coverage and the use of radar together with computers, meteorology in general has not kept pace with the advances that have taken place in the more sensational sciences.

We have seen astronomy moving into the realms of the radio telescope and with it a greater understanding of the universe around us.

Man in 10 short years had made one of the most sensational advances in space exploration by landing on and exploring sections of the Moon. The technical development necessary for that achievement is beyond the comprehension of ordinary man.

Exploratory probes have already been made towards the planets Venus, Mars and later further probes were made towards Jupiter followed by Saturn before heading onwards to the outer planets.

Man has opened his mind and pushed the old horizons in many fields further and further outwards. It seems that there is nothing beyond man's capabilities once he applies his mind to the problem at hand.

But the one unsolved problem man has not applied his mind to truly successfully is meteorology, that cinderella science controlled and kept within strict and confined limits by that isolated group known as meteorologists, a group which will not accept outside assistance, except only from its own persuasion. It is a group which has developed weather forecasting to its present level on but part of the knowledge necessary to be completely successful.

The other part of the knowledge necessary to a better understanding of the weather's idiosyncrasies has been completely ignored, mainly because its existence has not been recognised or admitted. The pathway to the other part of weather forecasting knowledge will be through an overgrown track in a jungle and this track will be tortuous and difficult to follow.

There will be many false trails and much back tracking before taking a broader trail which will lead to weather forecasts with an accuracy in outlook which is presently unknown. The rewards to the populations of the world will be no more wasteful planting before drought conditions, delayed harvesting, or over production because of excellent climatic conditions, or whatever.

Unfortunately, this new dream for sensible weather forecasting will not come from the present coterie of meteorologists, but from a new team who will have astronomy as part of their background knowledge, and past weather records, including weather maps, as their guide.

No more will the lecturer on meteorology continue to lecture on the outmoded ideas which have been perpetuated from the past and handed on to the later graduates. New skills will be taught to gain better and longer range weather forecasts.

From the very earliest days of man's history the weather has played an important part in his development. As seasons followed seasons man began to notice signs which told him what the weather could be for tomorrow. He progressed by attempting to anticipate the coming season.

He initiated sayings or quotes to show that his observations had meaning: 'Red sky in the morning shepherd's warning, red sky at night sailor's delight'; 'Rain before seven stops by eleven'.

In attempting to anticipate the next season's weather, the Maori in New Zealand had a saying that when the cabbage tree flowered early, it could be a dry summer, or if the ducks nested high, it would be a wet season.

Each and every quote will be based on centuries of observations. Each country will have similar sayings peculiar to that country, and although it cannot be said that every quote is 100 per cent accurate, there will be some sound common sense behind many of them. So along with other factors it is the intention, throughout this book, to make an attempt logically to explain the practical aspects of these various sayings and in the process show where the meteorological research services generally have missed a few of those turnings in the trail when attempting to seek the ultimate in perfect weather forecasts.

Forecasting subterfuge

Some of the present forecasting methods used in the meteorological services will be listed with an explanation on what effect this may have on the user of the weather forecasts. One is called 'persistence' forecasting, i.e. to persist with a forecast until a positive change in the weather pattern compels the meteorological services to change the forecast, which in turn will be persisted with until another obvious change is imminent.

In an attempt to convince the public further that the service is up with events, another subterfuge is used. For instance, if there should be rain which has not been forecast for that day, then the meteorological service will forecast that rain for the next day, even if it should be fine during the following day, in the belief that the public will think that the meteorological service was aware that the rain was due but it had arrived ahead of the forecast period. This is particularly noticeable when flood rains are not predicted. Or if a frost on a particular night was not forecast, then a frost is forecast for the next night, although the following night may be fine and mild. The suggestion offered is 'We were aware that the conditions were developing, but these conditions also arrived ahead of the forecast period'.

The present method for forecasting the weather can be likened to a dog following his master; the master being the weather, the dog the forecaster who never really gets ahead, but frequently seems to trail behind.

Do meteorologists question?

Under the present system, the meterological services world-wide attempt to anticipate the probable movement of the weather systems without really understanding why a system is moving in any given direction, or why its direction may change, or even its speed of movement. These reasons are unknown to the meteorological services, and yet there are many pointers that would lead the inquisitive forecaster or researcher to a more certain method for forecasting the weather, a method which will bring tomorrow's weather, or next month's weather, or even next year's weather, or any weather, closer to today. Tomorrow's weather can be brought to 'just around the corner', with this new understanding of the movement of the weather systems which will enable the forecaster to forecast the actual weather for tomorrow, so that one can believe what one hears.

No more will the evening's television weather forecast for tomorrow be changed in less than 12 hours, and then changed again six hours later. No more will a farmer cut hay on what appears to be a sound forecast for fine weather, only to find that one day later, after hours of work and costly dollars, his hay is lying ruined in the paddock.

Hectares of ground will have been rendered unproductive for about two months, because the meteorologists have never opened their minds, never queried the true reason on what causes the development and then the movement of the many repetitive weather systems which cross the country year in and year out, figuratively with the meteorologists heads in the clouds and with a wish for an undisturbed and unworried worklife followed by a sure and safe retirement.

So why is there a need for them to push beyond the realms of the accepted 'orthodox' meteorology? The present 'orthodox' method will be shown to be very much a dead horse having been flogged for about 60 or 70 years or more, but never really paying its way.

The reason that the present forecasting method has been allowed to continue beyond a sensible development stage is that the meteorological services generally feed on their own knowledge and would not, or did not want to accept or acknowledge help which may have been available from other groups or individuals. These groups or individuals, known in their various local areas as 'long range weather forecasters', follow personal methods which are, in their turn, a development from the previously mentioned folk law on weather sayings and developed to a stage where past happenings are projected into the future as long range weather forecasts.

In direct contrast the meteorological services started on a high note which was sensible and acceptable in any new science 60 or 70 years or more ago, but since that time, and being certain that there is nothing new under the sun, they have remained with the theory that the weather is controlled by 'hot air rising and the cold air takes its place — a theory taught in primary schools during the late 20s and probably before then. Added to this theory can be included, that trade winds also are caused by that combination of 'hot air rising and cold air replaces', together with the spins of the Earth on its axis coupled with the effect of the 'Coriolis' Force. Anything else which may happen, is accidental or incidental.

If meteorological services had recognised, encouraged and researched the methods of the long range weather forecaster instead of generally ignoring them and applying ridicule to their methods, the 'orthodox' weather forecasts of today could have been improved upon considerably if some of those methods had been added to the present forecasting methods.

The added use of radar, satellites and other modern aids have only marginally improved the weather forecasts of today, followed by a little more applied psychology directed at the public and a touch of ambiguity in the forecasts.

Most of the long range forecasters work without these modern aids, relying on past weather records, their interpretation of them and then applying them to future dates. Each may use a different system. All will have differences, all will not be correct, but each of them may have just one small factor worthy of intensive research. Each is paid a fee for his forecasts, so if they are useless he only sells once. If they are good he stays in business. Many have become legends in their own area, and that they have been thought of as such is a compliment to the long range weather forecaster, but it also shows dissatisfaction with the meteorological services world-wide.

But as this dissatisfaction is not generally highlighted by the user, the meteorological services can turn a deaf ear to any dissatisfaction which may surface from time to time and then just ignore it.

Meteorological services are generally autocratic, will not take criticism, and for this reason alone live in ivory towers. This attitude would preclude any serious attempt at researching long range weather forecasting based on the above suggestions.

The ideal atmospheric distribution

Sir Graham Sutton, a past Director of Meteorology in Britain, in his book *Understanding the Weather* published in 1960, explains the movement of the atmosphere with the explanation that warm air rises in the equatorial regions and descends in the region of the poles. This is referred to as 'starting up' the atmospheric engine, an action which, in turn, is related to the spin of the Earth on its axis to give us the wind direction in the various climatic zones.

This explanation is an oversimplification of the atmospheric movement under ideal conditions.

Under these conditions the weather should be exactly the same on any given day of the year, as compared to the same day in any other year.

The reason for this particular suggestion is that the Earth spins on its axis once every 24 hours — no more, no less. So there is no change here which will alter any atmospheric movement.

The Earth does one orbit of the Sun each year, so that the Sun's declination (its latitude) is almost exactly the same on any given day of the year compared to the same day of any other year. In general, the Sun pours the same amount of heat into our atmosphere each day, so our 'atmospheric engine' should respond by giving the same results year by year, season by season, month by month and day by day. The heat rising from our 'atmospheric engine' will be almost constant, so the direction and speed of the wind will also be constant when compared with the same day for any other year.

Frosts will be on the same nights. The same amount of cloud will form with rain on the same days and at the same time. There would be no early spring or late autumn.

The summers would be exactly the same as would be the other seasons of autumn, winter or spring.

Areas prone to droughts would continue to be drought areas. Wet areas would always be wet areas.

Anticyclones would be at the same intensity and on the same latitudes for the same period of time of each year. Depressions, if any, would have the same barometric pressure for the same period of each year.

This should be the expected pattern for the weather if the theory of the 'atmospheric engine' can be accepted as fact, and if the explanation by meteorologists is factual.

2: *What of gravity?*

But the previous comments overlook the one important natural law, that of gravity. If the law of gravity is to be believed — and it can be believed — then the gravitational attraction of the Sun towards Earth would draw more of the Earth's atmosphere to that side facing the Sun. This would create a high atmospheric tidal bulge with the Earth turning on its axis inside it. The heat of the Sun would, in turn, cause the moisture evaporated from land and sea to form a cloud on the side facing the Sun. As a result, that sunlight itself would not reach Earth, but some heat-generating rays would. The days would be continuously fully cloudy, the nights possibly clear and cold (probably freezing cold away from the tropics).

Under these conditions life as we know it could not exist. In fact it is possible that no life would ever exist on Earth.

The Earth with a Moon

Why then is it that none of these conditions exist as fact at the present time, or apparently at any other time in the history of Earth?

The answer lies with our satellite, the Moon. The Moon is the true disturber and distributor of our weather. The Moon is the reason for the changing intensity of anticyclones and of depressions. It is the explanation for the formation of hurricanes and also the force and direction of winds.

It is the reason for the changing temperatures and why the Arctic and Antarctic are where they are. The Arctic and Antarctic are the results of this weather, and not a storehouse for weather as most meteorologists would have us believe, although some of the cold from these areas will be redistributed outwards when the wind direction is favourable for this to occur.

Yes, that romantic and mysterious luminary of the heavens is a veritable Jekyll and Hyde. As Dr Jekyll the Moon can be benign and mellow with pleasant anticyclonic days and Indian summers, offering a blissful and pleasing lifestyle. It is the Moon that the romantic poets base their dreams on. But as Mr Hyde, the change is remarkable, with fierce storms, raging seas and wind-torn countryside, floods, frosts and droughts. Yes, even roaring volcanoes, tidal waves and earthquakes are all part of Mr Hyde's unpleasant characteristics.

Hard to believe?

Moon's orbit is irregular

The Moon is the only unstable factor in an otherwise stable situation. Without a moon, the Earth-Sun relationship would be stable even if the Earth was uninhabitable.

The facts on the unstable Moon-Earth relationship, are:

Our moon is large as moons go, when compared with the size of our planet. It also orbits very close to its host planet when its size is taken into consideration.

The diameter of the Moon is 3481km (2163 miles) and the Earth 3.6 times larger.

The Moon has an apogee, a point where the Moon is most distant from the Earth, and a perigee when the Moon is at its closest point, a difference of about 50,211km (31,200 miles). This difference between apogee and perigee is sufficient to add 20 per cent to a high sea tide when the Moon is at perigee, for this is when the Moon's gravitational attraction is at its strongest.

Apogee and perigee take place progressively as a precession on the Moon's orbit about Earth, i.e. in a different position on each orbit each month.

The plane of the Moon's orbit around Earth has this variable eccentricity which is slightly egg shaped. It also has a tilt or inclination of about five degrees from the plane of the Earth's orbit around the Sun (the ecliptic) which can be added or subtracted to the 23½ degree tilt of the Earth's axis (the 23½ degree tilt of the Earth which is the angle as viewed in relationship to the plane of Earth's orbit about the Sun). This will give a maximum declination (latitude) for the Moon of 28 degrees plus, north and south of the Equator each month, when the five degrees is added to the 23½ degree tilt of the Earth.

About nine years later, the declination of the Moon is about 18 degrees plus, north and south of the Equator when the five degree tilt, or inclination, of the Moon's orbit is subtracted from the 23½ degree tilt of the Earth. The full cycle from 28 degrees and back to the 28 degrees takes about 18.6 years to complete.

The maximum and minimum declinations of the Moon always take place at the Sun's solstice position over both hemispheres in its turn each month.

In the years between the 28 degree maximum and the minimum 18 degrees declination of the Moon, the amount of the declination is progressively reduced each year or, more accurately, each month. The position of the north part of the five degree inclination moves away from the solstice section of the orbit and steadily forward along the orbital path of the Moon. This reduces the declination progressively until the five degree north part of the inclination reaches a point 180 degrees away. It is now back to the solstice position. It is then the five degrees is subtracted from the 23½ degrees of the Earth's tilt and, at that time, the Moon's declination is reduced to 18 degrees, i.e. as it appears in relationship to the imaginary latitude lines on Earth — the Moon's inclination has now reversed itself and it is sloping in the opposite direction.

The perigee-apogee cycle, which also changes its position on the orbital path but moves in the opposite direction to the Moon's precession or changing declination, has a different timetable, taking 8.85 years for the complete cycle.

The Earth's tilt always faces in the same direction when it is orbiting the Sun and the Moon's maximum declination for that month, at whatever the number of degrees it may be, north and south of the Equator is always at that point which approximates with the summer or winter solstice, i.e. the northern declination of the Moon is the same point as the summer solstice in the Northern Hemisphere and the southern declination of the Moon occurs at the same point of the Southern Hemisphere's summer solstice.

But as a distinction; the north or south point of the Moon's five degree inclination will be at that part of the Moon's orbit where its position will be represented by the Moon's declination at that time. For instance when the Moon's five degree high point of its inclination is at one equinox position, i.e. sloping from above the ecliptic to the opposite equinox and below the ecliptic by five degrees, the declination movement of the Moon as seen on Earth will be about 23½ degrees north and south of the Equator, the same as the tilt of the Earth. This is because the axis of the Moon's tilt will be along the line of the ecliptic.

This situation remains the same each lunar month, irrespective of where the Earth is positioned in its orbit around the Sun. Also in each instance, the new moon of the summer always occurs near the time and place of each summer solstice, in either hemisphere, while the full moon is at the winter solstice for each hemisphere.

When the Earth is at either of the equinox positions, the new and the full moons take their places alternately over the Equator, which is where the equinox is positioned on the ecliptic (the plane of the Earth's orbit around the Sun), while the first and last quarters of the Moon take place at the summer or winter solstice positions and each alternating to the position of the other as the seasons change, as do the full and new moons at the spring and autumn equinox.

While the northern and southern declination of the Moon takes place at the same point of the Moon's orbit around the Earth, i.e. at the solstice position when related to Earth, the phases of the Moon progressively shift on their orbit around the Earth through the year, relating to the Earth's position, or angle, to the Sun in the Earth's orbit around the Sun.

Are there atmospheric tides?

Possibly since the time of Sir Isaac Newton, it has been known that the Moon is responsible for the sea tides and that when the Moon is new, the high tides rise higher and the low tides are lower, and like eclipses of the Moon, the rise and fall of the tides are predictable.

The spring tide can be extra high at different points on the Earth's surface, depending on the declination of the Moon, (its latitude) and the position of Earth in its orbit about the Sun — which is also the Sun's latitude north or south of the Equator — the result of the Earth's tilt on its axis, and of course the position of Earth on its axis, i.e. the time of the day. These various factors determine where the highest tides will be and when.

Although it is generally known and admitted that the Moon causes sea tides, that the height varies from place to place, and that all these factors can be predicted, it has never been admitted by meteorologists or astronomers that there are atmospheric tides with sufficient variation in height to influence the weather, or that the weather is but atmospheric tides, with variations the same as with the sea tides, having both high and low tides.

This may seem impossible when it is known that sea tides rise and fall twice a day, but anticyclones may last for as long as a week or more with the barometric pressures remaining more or less constant through that period, and depressions can be about an area for up to two weeks, possibly with very little variation in their barometric pressures also, and with pressures that can be extremely low at times.

How can the Moon 'hold' an anticyclone over an area for so long without any apparent sign of an atmospheric tide being noticeable, or a depression over an area for the same reason?

Meteorologists and scientists and many others have made one mistake since the barometer was discovered and became an important aid in meteorology. This instrument has been believed in implicitly and it was thought that further research was not necessary on the message it gave.

The high pressures of an anticyclone have been explained by stating that cooling air from the tropics becomes heavier when descending over the temperate zones thus giving the higher pressures. But what of low pressures? If descending cold air is said to cause the pressures to be high, then the opposite situation should apply with rising warm air giving the low pressures. Low pressures are usually associated with depressions and many, but not all, begin in the cold latitudes, so why the low pressures? Meteorological literature is strangely silent on this subject and never mentions the anomaly between the stated cause of the high pressures and the not explained low pressures; nor has there been an explanation as to why the air pressure changes in any case, for the atmosphere is there and its thickness should remain constant and with the same barometric pressures day in and day out. This we know is not so.

It can be said of meteorology as with other areas in science having problems, that a little knowledge is dangerous. In this instance the barometric pressure readings have been considered infallible.

Yes, the barometer does give atmospheric pressures, but not atmospheric height. These are not the same; pressures can remain constant day and night, but the atmosphere can alter in height twice a day as does a sea tide. This does not necessarily happen in unison for, whereas a sea tide will be restricted in its movement by drag and land masses, an atmospheric tide is not.

These atmospheric tides can vary in height by several miles and, unlike sea tides, cannot be observed — although there are many signs that show these tides do exist, and many sayings which, unwittingly, recognise their existence.

Gravitational attraction

The reason that atmospheric tides have not been discovered despite being extensive, was because the natural law of gravitational attraction, or the effects from the centrifugal force, was not applied to the atmosphere, whereby the Moon and the Sun individually, or together, depending on the phase of the moon, attracted the atmosphere towards either or both of them and, the important point is, they support some of the weight of the atmosphere by gravitational attraction which in turn does not show on the barometer.

Digressing a little and referring to sea tides, has anyone ever attempted to check if there could be any pressure differences between a high and low tide by placing a gauge say 100 feet down and then checking if the water weight has any unexplained discrepancy from that which would be expected between tides under normal circumstances?

So back to atmospheric tides; the Earth spins inside this atmospheric tide which will face towards the Sun or Moon, or both when they are on the same side of the Earth. The 'low atmospheric tide' usually has about the same barometric pressure as the 'high atmospheric tide' and although the height difference in the atmosphere can be considerable, this difference will not show on a barometer.

This atmospheric tidal effect and effects will be dealt with in more detail later, but one example at this point may suffice.

A new moon is a daytime moon and it is usually accompanied by daytime cloud during the middle of the day, even if it is only broken cloud. This is referring to anticyclonic weather, when the mornings are nearly always clear and the evenings will be mostly clear also. Temperatures are nearly always mild while the night skies will be mainly clear and cool even in the summer, this being the lower tide. These conditions depend on the position of the new moon on its orbital path. If the weather is unsettled, the rain will be mostly at night. Again this will depend on the position of the new moon and its latitude, which point will also be covered in more detail later.

The results from the centrifugal force, mentioned previously, will occur on the opposite side of the Earth and away from the Sun and Moon when at the new phase or during the first or last quarters, while the Sun and Moon are about 90 degrees apart. This will be when the atmosphere may be thrown outwards into a lesser atmospheric tide by the speed of the Earth's spin at the Equator. The air, being light, may not be thrown into as deep a tide as when the atmosphere is lifted gravitationally by the Sun or Moon and it seems likely the centrifugal force may not be a dominant factor in altering the depth of an atmospheric tide.

Any astronomer would know that that which has been written up to this point will be correct as far as the gravitation attraction theory is concerned.

At this stage it may be a good opportunity to digress again and apply the same thoughts and theories to the atmosphere on the other planets as they orbit the Sun.

As Venus has no moon, and Mars only lumps of rock as moons, which gravitational-wise, could not attract any possible atmosphere on Mars towards them, it must follow that any attraction of any atmosphere and cloud must be towards the Sun. It must also follow that the Sun would always shine on any atmosphere and cloud on any of the planets, and it would also follow that we on Earth would only see that side of a planet with its atmosphere and cloud facing towards us. The rest of the planet would be in darkness so we would never see the other side of say, Venus, which could be cloud-free. Because any probe sent to Venus will have to land on the side facing Earth so enabling the probe to relay its transmissions back here, we can never be sure what is on the other side and how it would really appear to us. And because Venus is closer to the Sun than Earth, the Sun would have a greater influence on the atmosphere of Venus than on the Earth's atmosphere. The Earth would have the same appearance if it were without a moon and as viewed from space.

Visual proof

Besides using the barometer for pressure readings at ground level, meteorologists use balloons to gain some knowledge of the upper atmosphere. These balloons carry instruments to measure pressures and temperatures in the upper atmosphere.

Anyone connected with aviation, and many other people, will know that a barometer and an altimeter operate from the same mechanical principle. Anyone connected with airport control will agree that any aircraft approaching an airport will ask for a barometric reading so that the altimeter can be reset to read height correctly for that particular area. Then the aircraft can be sure of its required height for its approach to that airport.

A barometer carried by a balloon would only be useful close to the area in which it was released. Once the balloon had drifted over another area with a lower or higher barometric ground reading, the information transmitted from the balloon may have very little value unless the true height of the balloon was known, for owing to the rise and fall of an atmospheric tide, the balloon would rise and fall with the tide at a buoyancy level in the same manner as a ship rises and falls or floats on a high or low tide at sea. The passengers and crew on a ship are not aware if the ship is on a high or a low tide.

In the same manner, the barometer carried by the balloon does not show at what height the balloon is floating. It is said that they burst at about 60,000 feet or could it be 50,000 feet or even 70,000 feet. It may be that any information collected by the balloon could be suspect.

Pilots of propeller-driven aircraft will have had the experience at one time or another of having difficulty in getting an aircraft up to a required height of say, 10,000 feet quickly.

On one type of day the plane will climb easily to 10,000 feet but on another day or even later in the same day, considerable effort and distance will be required for the plane to climb to the same height.

The pilot may consider that the motor is not functioning as efficiently as usual, or offer some other excuse as a slight down draft which may slow the rate of climb, when the real reason will be the variation in the atmospheric tide or grip on the air. On a high atmospheric tide an aircraft will climb readily to 10,000 feet, but on a low tide, the aircraft is struggling to climb to the same height, the air will not have the same 'body'.

With this thought in mind, some consideration could be given to aircraft accidents caused by misjudging the climbing rate of an aircraft on certain days, for there are days when the atmosphere can be very 'thin'.

Signs of a low atmospheric tide are with us throughout the year. We have all experienced those very hot days which are never completely cloudless, but do have long sunshine hours. The tar sealing on roads and footpaths bleed; temperatures are high and conditions are uncomfortable. Why?

Usually our atmosphere absorbs a fair proportion of the radiation emitted by the Sun as heat, but during the period of daytime low atmospheric tides, the atmosphere is not thick enough to absorb all those strong heat generating rays and so those rays penetrate to ground level and tar, being black, will absorb this extra radiation as heat and the tar will bleed.

A further phenomenon of these conditions will be the sudden cooling of the atmosphere when a cloud obscures the Sun. The cooling is positive and sudden giving further evidence on the lack of sustenance or thickness in the atmosphere which would normally hold the heat.

The reason for the coolness! Cold is the natural condition in space and space conditions are brought closer to the Earth when the atmospheric tide is low.

Also, this condition only occurs during certain phases of the Moon, generally when the Moon is full or in its last quarter. When the Moon is full it is night-time moon, which means that the Moon and the Sun are on opposite sides of the Earth.

The Moon appears to be the more dominant factor in giving a high tide; so when it is a night moon, the midday atmosphere is at a lower ebb. This in turn allows a greater input of solar energy resulting in warmer temperatures.

A further factor will depend on which hemisphere the full moon is over. During the winter months of the Southern Hemisphere, the full moon is over the Southern Hemisphere, then changing slowly to be full when the Moon is over the Equator at the spring equinox, followed by full over the Northern Hemisphere during the winter months there. So it will follow that while the Moon is full over the north during the southern summer months, the greater tidal bulge of the atmosphere will be transferred to the Northern Hemisphere, giving an extra low atmospheric tide over the south.

This midday low tide will be transferred slowly towards the later afternoon as the Moon moves towards the last quarter phase and then to the late afternoon by the last quarter.

During the last quarter, the Moon sets near midday so there will be no moon in the sky in the late afternoon or early evening and consequently a low atmospheric tide then.

The conditions of a low atmospheric tide during our summer months have been mentioned because this is the busiest time for summer sports and recreation. Some of the sports are energy consuming such as athletics and of these, the marathon or decathlon will be possibly more energy consuming than most, and it is this type of sport which is very prone to heat exhaustion.

A check on records will show that in many instances, collapsing from heat exhaustion will occur during the full moon, and then in the period up to the last quarter; a fact which should convince the promotors of sports, regattas and the like, to hold any event which may cause heat exhaustion amongst its contestants, in the cool of the early morning. Conversely a first quarter moon could provide similar problems with an early morning sporting event, although in this instance the Sun may not provide that excessive heat in the morning that it may provide with a last quarter moon in the afternoon.

The problems from a last quarter moon during the summer months will be very apparent to those who enjoy sunbathing. If it is your desire not to be toasted to a cinder, don't sunbathe for long periods at this time of the month, especially if it should be your first sunbathing of the summer, even half an hour may be too long, or one hour far too long. You have been warned.

There are many signs of a low atmospheric tide, and many of them are observable and recognisable once the phenomenon has been mentioned. It can be very hot and the daytime temperatures will hit all-time highs for the summer.

In the winter all-time low day temperatures will be likely, particularly if the wind is from the cold quarter, south-west to south-east in the Southern Hemisphere and north-west to north-east in the Northern Hemisphere.

In countries prone to heavy snow storms or blizzards, check the quarter of the moon when they occur. Most likely it will be full or the last quarter. If the moon is new, the snow will be, in most instances, at night because the Moon is a day moon and in the winter it is over the other hemisphere. Or if it is at first quarter, the snow will be at night and into the morning.

A general rule on if it will be fine or not, whatever the season, is; if the Moon is up in the sky it will be less likely to rain, but without a moon in the sky, rain, if about, will be more likely.

This comment will apply in the temperate zones and as the author has not had any experience on conditions in the tropics or in the high latitude zones, it cannot be guaranteed that these comments will apply in those areas.

Another sign of a low atmospheric tide will show with chromium plating on motor vehicles etc. glinting brighter in the sunlight and also from the windscreens or from rear windows. This extra brightness can be very annoying to other vehicle drivers although it must be admitted nothing can be done about this problem.

This brightness can be observed on water when sunlight is reflected back.

Meteorologists in particular, and scientists in general, have always insisted that the atmospheric tide is infinitesimal and that it is of little consequence in its effect on the day to day or seasonal weather, or even climatic changes.

Proving atmospheric tides

The problem in attempting to prove that there is an atmospheric tide was of considerable concern to the author, for theory was one thing but positive proof another. Very few would admit to the probability that tides in fact do exist, and not only exist but could be substantial in their variation from low to high.

It seemed obvious that if a tide in fact did exist, it had to be very mobile.

Because of the changing position of the Moon in relation to the Sun, i.e. its phases, which also are always related to the Earth's position on its orbit around the Sun, it then meant that the Moon and Sun would combine their respective gravitational attractions say, when at new moon. When both are positioned over one hemisphere (as a new moon), and when both the Sun and the Moon are at their maximum declination (latitude) either north or south of the Equator, they thereby attract a greater percentage of the atmosphere to that part of the Earth immediately below the Sun and Moon.

Then there are the different angles of the Earth-Sun-Moon when the Moon's phase is in either quarter, first or last, i.e. at right angles to the Sun. The Moon's quarters can occur on any part of its orbit around the Earth which will depend on what the season is at that time. This in turn means that each phase will occur in a similar position each year, but not on the same date, in relation to the Earth's position on its orbit around the Sun. The changes in position apply equally to a new moon or a full moon and all angles in between.

So the question was, how could one approach the problem of proving that there may be a massive tidal change in the atmosphere. One cannot walk about on top of the atmosphere. Flying would not be of any assistance. A balloon may help, but to be useful it would need to be anchored so that the changing atmospheric tide would pass over the balloon. The anchor rope would require a measuring device incorporated in it to measure any extra pressure applied to the rope as the balloon attempted to rise higher with the rising atmospheric tide, for the balloon would attempt to rise to a new buoyancy level which the higher atmospheric tide would create.

This would be a satisfactory method if the conditions for the experiment were ideal, no wind to offer that extra pressure on the balloon and no rain to add further weight. So it would seem that testing the 'ideal atmospheric model' would not be possible using that method.

After much thought, the author decided that the best approach would be to use a meter which would give a solar reading on brightness. A photographic exposure meter was obtained and filtered to a degree that when the meter was faced towards the Sun, a reading was obtained. The face of the meter was graduated to give a reading which read at 14.0 and up to 14.6 or 7. A telescope type level with degree markings from the horizontal and towards the vertical was also used, so that the angle of the Sun could be measured, and this angle was entered with the brightness reading.

After a while it soon became evident that cloud too close to the Sun would affect the reading, so all readings were then taken only on suitable occasions. The readings were taken at any part of the day, from early morning to late in the afternoon, so that all Sun angles were catered for, and also seasonal changes could be noted together with the various Moon phases.

It is acknowledged that the equipment will be considered 'Heath Robinson' but it was sufficiently accurate to establish a pattern.

Barometric pressures were available to establish if any changes in brightness could be related to them.

The extent of atmospheric tides

The brightness readings under similar conditions, but different Moon phases, varied by an amount which suggested the atmospheric tide could alter by as much as 25 per cent. So if one considered the useful atmosphere to be five miles thick, then the atmospheric tide may alter by as much as one and a quarter miles, or if the accepted total depth of the atmosphere at about 60 miles is used, the tidal differences between a low and a high atmospheric tide could be of the magnitude of 15 miles. It must be said that these figures are maximum and relate to the upper levels of the atmosphere where the air is very thin.

Some confirmation for these figures could come from say NASA for when spacecraft re-enter the atmosphere, some changes must be noticed each time a craft returns from space. The re-entry would not necessarily occur at the same quarter of the Moon so from this fact it may be possible to establish if a craft 'skims' on the upper atmosphere for longer or a lesser period before gliding into the lower atmosphere. It may be possible that a spacecraft on entering the upper atmosphere could commence skimming at different heights depending on the phase of the Moon, if this is so it could offer some confirmation in part that atmospheric tides in fact do exist.

The meter for brightness brought forward another phenomenon; readings always increased in intensity before rain arrived and rain could be predicted for the next day almost with certainty. In most instances the brighter readings related to that part of the day when rain could be expected.

Also the reading would increase while the barometer remained high, but with a frontal system approaching.

If a day was completely clear and the meter readings were high, the following day nearly always was completely cloudy.

It could be argued that the phenomenon of higher readings was the result of high level moisture in suspension in the upper atmosphere acting as a reflector thereby increasing the solar readings. This could be true, but as humidity readings were not taken, no comparison on this point was possible and also, if there were some humidity, one would expect the sky to be slightly hazy and not clear.

In case some criticism may be offered on the type of equipment used in the experiments, and which could be considered archaic, it must be said the results were quoted because of the length of time this equipment was used. In more recent times a Cushing solar energy meter has been used to verify the previous results. This meter has been used only when the sky was mostly clear and for this reason the opportunity to use the meter has been limited, and the results have been slower to gather. However, much the same picture has emerged and it shows a similar percentage of change in the atmospheric tides.

The author is aware that solar readings are taken for a number of reasons — for ultra violet, infra red etc. — and it may be that if some of these readings have been recorded on a graph, and if some research is undertaken on these graphs, then confirmation or denial may be possible.

This then brings forward the thought that any solar radiation readings taken previously for whatever reason could be suspect if they have been affected by a changing atmospheric tide.

If the proposition is that the very high atmospheric tides do exist, then the knowledge could be useful for mountaineers when attempting to climb very high peaks, for it would be obvious that the better results, especially without oxygen, would be possible if climbing were undertaken while the atmospheric tide was high, and then resting when it was low. The best climbing time would be when the Moon was new and when the Sun and Moon were near the same latitudes where the climb was to be undertaken. This would be a daytime high tide which would be at its highest in the area of the climb, and of course the atmospheric low tide would occur during the night hours and while resting.

3: Moon phases and the weather

Over the years the author has accumulated many newspaper clippings on extreme weather conditions of varying kinds, such as tornadoes, hurricanes, electrical storms, snowstorms, hail in excessive amounts and size, heat waves, or any other manifestation of weather normally referred to as 'Acts of God'.

But let it be said immediately that as far as the weather is concerned, there is no 'Act of God' in the sense that God does actually control the weather, and He will alter it at will. If there were to be any divine intervention, it would have been a programming at the beginning of creation, and not something that is alterable at will.

God will not provide a fine day because a group of people selfishly pray for such a day, nor will mass prayer end a drought, because all the elements which make up weather are there and they are unalterable until the normal span for such weather has passed and a new phase begins.

Having made that point, it can be said that there is a time for the variously listed events.

For instance tornadoes will occur under certain conditions which are dictated by the phase of the Moon. Whenever a tornado has been reported in the newspaper, and a time has been mentioned, a check will show that the Moon has still to rise or it has set. This situation applies, even if the tornado is at night, the early morning, midday or late afternoon.

A tornado at night usually occurs during the new moon, which is a day moon. If the tornado is during the early morning hours and up to about midday, the Moon will be more than likely in its first quarter. But most tornadoes occur while the Moon is full, and into the last quarter for this is the period when the Sun applies the most heat to the ground, which creates the perfect conditions for tornado activity. There may be odd exceptions, and that is just what they are, exceptions to a general rule.

Meteorologists frequently say that there is no proof that the phases of the Moon affect the weather.

They should open their minds to positive facts.

A file on hurricanes, cyclones, typhoons or whatever name one cares to use, will show a similar preference for lunar influence, but because these destructive storms can take a number of days to develop, the overlap into other phases of the Moon not normally associated with this phenomenon, will take place in a limited way.

By far the largest file on these storms shows that the majority occur from the period of a full moon and into the new moon period, and with most of the emphasis on the last quarter.

A considerably reduced number take place during the first quarter of the Moon.

A further breakdown on the occurrence of hurricanes will be dealt with in another chapter, and also reference will be made on the seasonal character of the hurricanes as well.

A cloudburst

The weather phenomenon commonly referred to as a 'cloudburst' is also moon phase based.

This is referring to a localised period of exceptionally heavy rain as distinct to heavy rain caused by two opposing fronts meeting; one cold, the other warm.

A cloudburst can be responsible for a flash flood which can cause considerable damage and even loss of life should one occur near a populated area and close to steep hills with gullies facing towards towns or cities. The considerable quantity of water generated by the cloudburst can cause even small streams to rise suddenly and to a height where a wall of water will crash into buildings and through streets flattening or damaging everything in its path.

Before a cloud can hold sufficient moisture to become a candidate to 'burst', an updraft is usually necessary.

Raindrops fall at a maximum speed of 27.36km per hour (17mph); a higher speed will cause the raindrops to break down and form a mist.

If an updraft is ascending at the same speed as the raindrops are attempting to fall, the raindrops will become stationary or nearly so and the cloud could continue to accumulate further moisture until it becomes super saturated.

Such a cloud is easily recognisable as it will stay almost stationary over an area and will have that ominous dark appearance.

The updraft speed will be the all important factor in determining whether the cloud will develop into one which will 'burst', for if the updraft speed is above the 27kph, the raindrops will rise higher and form hail in the colder air above and so remove the danger that the cloud will burst. When heavy enough, the hail will fall to the ground.

The type of cloud which will 'burst' or form hail is the cumulo-nimbus variety and this cloud will form only when the Moon is full or into the last quarter.

When the updraft ceases through loss of ground heat and coolness from space brought about by the lower atmospheric tide, the cooled super-saturated cloud will be unable to hold its excess moisture and drop its contents in one short fall.

A cloud prone to 'burst' is difficult for meteorologists to predict and about the only warning that they may be about to develop will be from the cloud cover increasing during the late afternoon on successive days and showing that ominous dark appearance. Otherwise conditions are usually settled looking and with a stable barometer so that there will be nothing to indicate that rain is imminent. For this reason clouds that will 'burst' are seldom predicted.

When will hail form?

Hail storms also nearly always occur while there is no moon in the sky. The exceptions again are minor, or hail may occur in high latitudes while the Moon is positioned over the opposite hemisphere — positioning which creates a lower atmospheric tide in the area affected by hail.

Hail is usually associated with towering cumulo-nimbus clouds which reach into the troposphere, 10 miles high where the temperatures are below the freezing point. Cumulo-nimbus clouds are unstable with strong updrafts which cause rising moisture to form hail at the higher altitudes, and after gaining weight and then falling to lower levels, the hail will be repeatedly carried aloft into the freezing temperatures where the extra moisture collected at the lower altitudes will freeze further and form larger hailstones.

Because hail occurs at the times mentioned, it would be logical to assume that the height of the troposphere must be variable or that the freezing level in the atmosphere must be brought down to a lesser height through the Moon's tidal effect on the atmosphere; or to reverse the statement, when there is a moon overhead in the sky, the freezing level is at a higher altitude, which is too high for raindrops to be carried to in the updraft, so gravity will take over before the raindrops freeze, or if they do freeze then the hail will melt again into raindrops which will fall as normal rain.

This is usually the explanation offered by meteorologists under these circumstances when hail fails to reach the ground. Part of that explanation could be correct but not all of it.

It should not be too difficult to check if cooler temperatures from the upper atmosphere do descend to lower levels at a time when an atmospheric tide is low. Sufficient temperature records of the upper atmosphere should be on hand now, and all that will be required is to relate them to the phases of the Moon.

Other Moon-induced phenomena

Whirlwinds and a similar phenomenon, water spouts, also follow a moon phase orientation, but as these phenomena require rising heat for their formation, they will always occur when the temperatures are high whether on land or sea.

So it will follow that whirlwinds and water spouts form during the summer and into the warm autumn months, and nearly always when the Moon is close to or at full, and into the last quarter. It must be remembered also that a summer full moon and the last quarter will take place over the opposite hemisphere, which means that the Moon will attract a greater amount of the atmosphere towards that hemisphere, and so create a lower atmospheric tide in the hemisphere, which is experiencing summer or autumn.

These are the conditions conducive to whirlwind and water spout activity.

Heat waves will also follow the same lunar phases, beginning just before a full moon during summer or autumn. Given the right atmospheric conditions, such as a prolonged anticyclonic spell, the temperatures will continue to rise day by day until a full-blown heat wave is established. These conditions will continue until the Moon moves towards the new moon phase and because the new moon is a day moon and over the hemisphere with the summer season, the atmospheric tide will be higher during the day, thus screening out some of the heat from the sun in most instances. However, when a new moon is at apogee, the gravitational effect may be reduced by a fair percentage because the Moon is further away from the Earth and therefore an atmospheric tide will not be quite so deep which may allow a heat wave to continue for a longer period.

Another factor which could be mentioned in conjunction with heat waves is that although atmospheric conditions may be generally anticyclonic and possibly stationary or slow moving, the barometric pressures are also lower in most instances than that which is usually associated with anticyclones.

This suggests that the height of the atmosphere may be less when the Moon is at apogee, and so allowing a greater input of heat producing solar energy.

Meteorologists refer to the regions where anticyclones are positioned as that part of the globe where descending air arriving from the equatorial regions after it was heated there and on cooling, descends to form high pressure areas which are called anticyclones. But no explanation has been offered on why anticyclones have such a large range in their barometric pressures. One would suppose that descending air is just that, and with common pressures at the centre, but this is not so.

It may be argued that an anticyclone with warmer temperatures would be more buoyant and thus the barometric pressures would be lower. This also implies that the warmer air would rise to greater heights, which in turn would assist in screening out the heating energy from the Sun. But we do know that excessive heat does reach the surface under the conditions mentioned, and it must follow that the height of the atmosphere cannot be sufficient to screen out the heating rays — in other words a low atmospheric tide.

Perigee — its increased effect on the atmosphere

With sea tides, it is known that the height of the tides are influenced by the distance that the Moon is from the Earth. This is a variable factor and when the Moon is at its closest approach to Earth, it is then at perigee, and when at its maximum distance it is at apogee.

The position of perigee and apogee on the Moon's orbit is also variable, and does occur progressively in different positions on the Moon's orbit around the Earth within a cycle of 8.85 years. This means that perigee will occur when the Moon is over the Southern Hemisphere for about 4.4 years, and through all moon phases occurring during that time. Perigee will then move into the Northern Hemisphere for the remainder of the 8.85 year cycle.

Also, perigee and apogee are not always exactly 180 degrees apart. Apogee progresses almost steadily around the orbit but perigee alters its place on the orbit. It is attracted or influenced by an advancing new or full moon so that the angle between apogee and perigee will be less than 180 degrees and at other times it will go beyond the 180 degree angle as viewed when perigee is advancing around the orbit but continuing to move in the same direction.

When perigee coincides with the Moon near full, a night moon, or at anytime into the last quarter, the extra gravitational attraction on the atmosphere, created by the Moon at perigee, will give extra low atmospheric tides during the daylight hours near midday, and with the last quarter moon rising about midnight, during the late afternoon hours. If at such a time the season of the year is summer or into autumn, and the weather dry, then very warm temperatures can be expected, the days with that brassy look and with heat as if coming directly from a super-heated oven. This would be a time to stay indoors and avoid the possibility of being affected by a heat stroke.

It is also the time to remember that animals can suffer from heat as well, so provide water and shade for them.

Lightning — its origin

Electrical storms also follow the tendency to be about when there is no moon in the sky, whether it be day or night, and again the most positive time for electrical storms is during the last quarter of the Moon.

Another factor which can alter the general rule is the position of the Moon other than its phase. Writing of the Southern Hemisphere, especially over the higher latitudes, i.e. nearer the South Pole; if the Moon is at its most northern declination and at whatever phase, then electrical storms may occur with a moon up in the sky. But this is a minor exception considering the frequency that electrical storms do occur with the Moon out of sight over the horizon.

When the lightning is at night, the Moon usually will be at the new phase, a day moon, or into the first quarter when it sets about midnight so there is little or no night moon at this time.

These facts pose an interesting question, and the author cannot claim to be completely knowledgeable on the magnetic or electrical interaction between the Sun and the Earth or its effect on the Van Allen Belt.

That the Sun does bombard the Van Allen Belt (lines of magnetic force circling the Earth from pole to pole) with particles or ions from within the solar wind, is well known, and it is also well known that the Van Allen Belt acts as a screen against the Earth receiving an overdose of ions. Now could it be possible that the Van Allen Belt is also affected by the phases of the Moon, so that when the Moon is, say at full — a night moon — the Van Allen Belt is drawn in towards the Earth during the day, and the charged particles of the belt are then brought in closer to the Earth's atmosphere, which in turn becomes highly electrified, so providing the mechanics for violent electrical storms?

Now that the point has been raised, research will quickly confirm or deny this suggestion, and even if confirmation should be forthcoming, it will be doubtful that any action could be taken to prevent the occurrence of electrical storms.

Aurora Australis

Meteorologists, climatologists and other sciences have often speculated on a possible connection between the auroras (Aurora Borealis of the Northern Hemisphere and the Aurora Australis of the Southern Hemisphere) and weather.

In an attempt to obtain some indication on whether there may be some effect on the weather in New Zealand, 36 aurora sightings from areas within the country mostly around the south of the South Island, and those reported from southern latitudes shipping in the near vicinity of New Zealand, were obtained and associated with the prevailing weather systems at the time of the sightings. There were 24 when conditions were anticyclonic, eight when depressions were over or relatively close to the country and the balance of four with cold fronts crossing New Zealand.

As cold fronts are associated with depressions, they can be added to the depression numbers which will bring that total to 12. This figure is half of the number of those sighted when conditions were anticyclonic at the time the auroras were sighted.

Over a greater number of samples, the figures could even out, so it cannot be presumed that one type of weather will be more prominent than another, and also auroras, most likely, would be more visible during anticyclonic weather when cloud would normally be less than during unsettled conditions.

However, the main interest was in the Moon's phase at the time of the aurora sightings, and it was here that the analysis gave the greatest differences. The Moon was at or near new, a day moon, on 27 occasions, in the first quarter four times and the last quarter five times.

In the quarter phases in most, if not on all occasions, the Moon was still to rise, or it had set. This seems to indicate that the Moon could have some influence on when the auroras will be visible.

There could be an argument in favour of saying that a bright moon such as a full night moon could inhibit the sightings of an aurora. This may be so to a degree but the brightness of many auroras is usually sufficiently bright for many of them to be seen still, even if they are less spectacular.

Another interesting analysis was obtained from the position of the Moon at the times of the sightings. In 24 sightings, the Moon was over the Northern Hemisphere and generally still moving northwards. In four of them the Moon was over the Southern Hemisphere but also going towards the north and in the remainder (eight), the Moon was over the south in varying degrees of declination going south.

The intensity of the aurora displays has been linked to the sunspot cycle of approximately 11 years, and when the sunspot activity is low so are the aurora displays at a low ebb or non existent.

However, the main question at issue is: does the Moon have the same gravitational effect on the Van Allen Belt, which may be the cause of lightning, as on the charged particles that form the aurora's display. These are the same particles, as mentioned, which escape from the Van Allen Belt to form the aurora display, and it may well be that the particles will only escape into the ionosphere, from whence the auroras are visible, when the Van Allen Belt has been gravitationally drawn in closer to the Earth.

4: *The Coriolis Force*

One very important plank in the theory on meteorological phenomena is the 'Coriolis Force' named after a brilliant mathematician Gaspard Gustave de Coriolis, who published a theory about 1833 detailing the motion of bodies set in motion on a spinning surface. He proved that any body set in motion on a rotating surface curved in relation to any other object on the spinning surface, and it did not matter in which direction or where the object was set in motion.

There does not appear to be any evidence that Coriolis had the atmosphere in mind when he developed his theory. It was a purely mechanical theory, for he was first and foremost an engineer. Coriolis assumed the chair of mechanics at Ecole Centrale des et Manufactures in France in 1829.

However, an American named William Farrel, a school teacher studying wind movement, used the Coriolis theory to explain the reason why the wind moved in a particular direction in each hemisphere.

The 'Coriolis Force' as adopted by meteorologists, in effect, says that any movement of the atmosphere in the temperate zones will slowly turn to the right in the Northern Hemisphere and to the left in the Southern Hemisphere, i.e. to the east as viewed from the Equator, thus anticyclones circulate in a clockwise movement over the Northern Hemisphere and anti-clockwise over the Southern Hemisphere.

The other weather systems are also affected by this action.

Coriolis Force — on a disc

When Coriolis developed his theory, he probably had a disc in mind and not a sphere, because in a mechanical sense a sphere would have very little application in the field of mechanics but a disc, in whatever form, would be used extensively even during Coriolis's lifetime, be it as a wheel, gear or friction plate, or the many other uses to which a disc may be applied.

The theory applied to a disc would have an apparently fast moving centre where the circumference is small, so that any object on the disc set moving away in a straight line from near the centre and towards the outer larger circumference, would slowly arc either left or right, depending on the direction of the disc's rotation. This arcing is caused because the object, when on a shorter circumference near the centre, takes the rotational speed of that circumference and transfers it to the larger circumference of the outer circles as they get progressively larger and the distance of the circumference becomes longer.

If the distance covered on the inner circumference nearer the centre of the disk as it is rotating, is say 5cm, and the object is over that line, then that 5cm will be the distance the object will cover on each successive increasing circumference as it moves outwards, so that the object must fall behind in relation to the point where the object commenced its outward journey, thus an arc is formed.

Coriolis Force — on a sphere

On a revolving sphere such as the Earth, an object or the atmosphere will move out at right angles from the largest circumference, the Equator, and as the circumference of the latitudes become smaller, the atmosphere will tend to arc towards the direction of the rotation of the sphere. In the case of the Earth which rotates towards the direction we call east, the arcing is towards the east; in other words, the atmosphere apparently gains speed along the latitude zones as the atmosphere moves towards the poles. This movement explains why the weather systems advance from the west and towards the east, in general terms, over the temperate latitudes.

Trade winds

Because the Coriolis Force is not effective near the Equator, the direction from which trade winds blow (north-east in the Northern Hemisphere and south-east in the Southern Hemisphere) is explained by saying that the rotation of the Earth at the Equator causes a retarding action on the atmosphere, so that the winds there angle towards the Equator.

This explanation does not give a reason as to why the atmosphere is tardy by not following the Earth exactly, while spinning on its axis once each day.

The atmosphere is part of the Earth, so why does it not keep pace with it? The Coriolis Force offers a reason for the atmosphere moving in the opposite direction to the trade winds at higher latitude, so why not the same tendency nearer the Equator, even if it is only a slight easterly drift.

Perhaps the reason for the retrograde movement of the atmosphere near the Equator may lie in an as yet unstated and an undreamed of approach to atmospheric physics (an approach where a model atmosphere could offer sensible answers instead of the past answers which lamely finish when, after studying a model atmosphere and doing an analysis by computer, they then state that there is such a lot we do not understand about atmospheric movement). How often does one read a complete and explicit explanation on the behaviour of the atmosphere where everything is placed in its correct category and then have the article finish on a negative note — we have such a lot to learn?

A point has been made previously on the possibility that the Moon has a definite gravitational attraction on the Earth's atmosphere. The Sun also has an influence on the atmosphere.

We know that the Earth spins on its axis towards the direction we call east, which is the reason why the Sun rises in the east and sets in the west.

The Moon orbits Earth in an easterly direction, but because Earth's spin to the east is at a faster pace than the Moon's orbital speed around the Earth, the Moon appears to travel from east to west.

The combined gravitational effect of the Moon and the Sun will be that they will create a drag on the atmosphere, so that as the Sun and the Moon appear to travel westwards each day, this drag will cause a westerly drift in the atmosphere between the tropics and the Equator by means of this gravitational attraction, which in turn could produce the winds known as 'trade winds'.

The orthodox explanation for the trade winds is given as warm air rising at the Equator to be replaced by cooler air coming in at sea level forming what is known as a 'Hadley cell'. This replacement air is deflected by the Earth's spin which blows westward as trade winds.

The gravitational attraction of the Sun and Moon causing the atmosphere to move in a retrograde manner, seems a more logical explanation.

The doldrums

In between the trade winds of the Northern Hemisphere and those of the south, there is a band of virtually windless air called the 'doldrums'.

The doldrums are not steady in any one place, but move north of the Equator during the northern summer and south in the southern summer. The reason offered is that the doldrums follow the heat of summer, but could it be that the doldrum band of air is controlled by the gravitational attraction of the Sun? And as the doldrums are not static, could it be that even the summer position in each hemisphere may be influenced by the Moon's monthly shift between both hemispheres?

There has not been available a dated reference to a shift within the doldrums during a particular summer, but if there was a shift it may not be considered to be of any particular consequence. The shift, if any, may be related to a possible change in the temperatures at the time and which may not be necessarily the correct answer.

One comment, on crossing through the doldrums by sailing craft, did mention that the trade winds of each hemisphere, on occasions, did move closer to one another, thereby reducing the influence of the doldrum zone, and making the crossing, by sail, quicker and easier. Perhaps this could be lunar intervention from gravitational attraction. Only investigation will give the correct answer.

Drift by anticyclones

In the temperate zones of each hemisphere, the anticyclones and depressions drift in a generally easterly direction. The anticyclones almost always do so and the depressions drift either east or south-east over the Southern Hemisphere and east or north-east in the Northern Hemisphere and the reason given is that this is the zone where the Coriolis Force begins to exert its influence on the atmosphere and where the eastward acceleration is most noticeable.

As mentioned previously, the Moon orbits the Earth in an easterly direction, and its daily advance in its orbit around the Earth is variable, depending on whether the Moon is at perigee (closest point to Earth) or if it is at apogee (furthest point away from Earth).

When the Moon is at perigee, its orbital speed is faster than when it is at apogee, and the Moon will cover about 14.73 degrees in its orbit each day, while at apogee the distance covered is about 11.85 degrees or 2.88 degrees less. The degrees covered reduce each day, decreasing from the 14.73 degrees of perigee to the minimum 11.85 degrees when at apogee, and increasing daily again to complete the orbit. The average daily motion is about 13.29 degrees.

When this movement is related to the drift of the anticyclones, it will be seen that there is some accord. There are exceptions, but overall a case can be made for some similarity in the Moon's advance on its orbit and the eastward drift of the anticyclones.

Some allowance will need to be made for the difference in land distances in kilometres, in a degree near the Equator and a degree at say 45 degrees north or south of the Equator, where the longitudinal lines are closer together and therefore fewer kilometres per degree.

The Coriolis Force — is it consistent?

When considering the Coriolis theory on weather system distribution, or the possible formation of weather systems from the Coriolis effect, the main point to consider is: why is it not consistent each day and offering daily predictable movement of the atmosphere? Is there something wrong with the theory, or is there a flaw in its adaptation from a mechanical disc theory to a theory explaining atmospheric movement on a sphere? Can an excuse be offered because the theory deals with air and not some solid mechanical component?

Air has pressure, and it should be sufficiently compact to follow the law of the Coriolis Force, but it does not. If it did so, anticyclones would remain at the same latitudes and also be the same size, and one should follow another without a break, separated only by small cold fronts to allow for the friction of the opposing wind direction between the western edge of the leading anticyclone and the eastern edge of the one following.

Because the Coriolis Force should exert equal pressure from the Equator outwards over each hemisphere, then anticyclones etc. should form on the same longitudes on both sides of the Equator at the same time which means that the weather systems should be exactly the same and differing only in their temperature range as the Sun moves from a summer season to a winter season over each hemisphere. But we do know that this does not happen.

In the tropics, the trade winds blow steadily at the same speed and cyclones, hurricanes or typhoons would be unknown because the steadily blowing trade winds would disperse the warm air and the sea water would not warm sufficiently to a temperature which is necessary for the formation of these destructive storms.

The wind direction and speed, in the temperate zones, would be constant altering (within the anticyclones) only as the anticyclones drifted steadily eastward.

The storms of the high latitudes would remain there.

The seasons would be predictable and stable, changing only as the Sun moved between the hemispheres.

For, remember the Coriolis Force is a continuously thrusting outward pressure on the atmosphere from the Equator and towards the poles.

The Earth's spin is steady and constant. It orbits the Sun steadily and its orbital speed has only minor variations, and these occur in the same place in its orbit each year; so again everything should be predictable.

So why do we not receive that regularity in weather patterns which all that consistency should give us? All the factors, on which meteorologists would agree, that give us our weather are there, but it is not as it should be — consistent.

What could be the reason? Could the Moon exercise more influence on the atmosphere than previously thought?

Every now and again someone has raised the question on the possible lunar effect on the weather, and meteorologists, mostly, have been adamant that there is no influence from that quarter. But why not?

An early attempt at proof

It is known the Moon has a positive and definite effect on the sea tides.

French mathematician Pierre Laplace explained the motion of the tides in 1773. He also wondered if there could be a similar tidal effect in the atmosphere. So he kept barometric pressure readings, taken four times a day for eight years, and found a rise and fall in pressures in each 12 hours, caused by the Sun's gravitational attraction, but no sign of a lunar influence. Laplace could not offer an explanation for this apparent anomaly.

In 1953 a strange fact was established by scientists; solar tides in the atmosphere were 16 times more powerful than lunar tides which was in conflict with the theory at that time. But in more recent times with the development of rocket research, a theory now suggests that the Sun tide is caused by the absorption of ultra-violet rays in the ozone layer 32-48km (20-30 miles) above the Earth, causing expansion and a hump facing the Sun. But this explanation does not offer a reason for a similar tide at midnight when the Sun is not shining nor does it explain how the expansion of the atmosphere can increase its pressure.

Anyone possessing a barograph with its continuous record of barometric pressures will observe that on most days and nights there is a definite lift in the pressures near midday and midnight. The exception occurs when the pressure is reducing. A question on that point could be asked, why isn't the ozone layer expanding and cancelling out the drop in pressures at that time? The ozone layer is above the normal cloud level so the ultra-violet rays should still function.

But conversely, a barograph usually rises at a faster rate on a rising barometer near the midday solar tide. The rise with the solar tide commences about mid-morning and drops to a level about 3pm.

The only explanation for the solar tides and higher barometric pressures could be that while the Sun does create an atmospheric tide by gravitational attraction at midday, the gravitational attraction is not sufficiently strong to hold the extra atmospheric depth within the solar tide without the barometric pressures increasing. The midnight rise in barometric pressures could be in reverse, in that the atmosphere is slack and without support from gravitational attraction, and so the true atmospheric pressures will show on a barograph because of the higher natural pressures.

This situation could suggest that the midnight tide may not have the atmospheric depth of the midday solar tide as the greater bulge or depth of the atmosphere will be on the side facing the Sun but not all the atmosphere will be supported by gravitational attraction.

This comment has been made without bringing into consideration any influence the Moon may have on the atmosphere during the middle of the day or at midnight.

Another view which may need to be considered may arise from the same reasons which create the two high sea tides each 24 hours. One high tide follows the Moon and it is caused by the gravitational attraction of the sea towards the Moon while the second high tide, 12 hours later, occurs when the Moon is on the opposite side of the Earth.

This tide is caused by the opposite force to gravity which is the centrifugal force whereby the spin of the Earth on its axis has a speed, at the Equator, of 1609kmph (1000mph). At this speed the centrifugal force attempts to throw the sea off the face of the Earth but gravity being the stronger force of the two, still has sufficient strength to prevent that occurring. However, the centrifugal force is still strong enough to create a high tide on the opposite side of the Earth from the Moon-created high sea tide.

The same forces which give us the two sea tides each day must have a similar effect on the atmosphere. Just what the effect on the atmosphere will be one cannot accurately predict because of the various phases of the Moon which occur on any part of its orbit during the year. It also follows that the phases of the Moon will take place over any latitude (declination) within the Moon's shift north and south of the Equator over each hemisphere. So the atmospheric high or low tides will similarly take place wherever and however the Moon is placed.

Although Laplace could not identify a lunar atmospheric tide, they do exist, and the mistake he made has been perpetrated for over 200 years.

He and meteorologists ever since, have overlooked the fact that air, unlike water, can be compressed to higher pressures and air can also be expanded to give lower pressures. The expanded air with the lower pressures at times will have a greater atmospheric depth than compressed air with higher barometric pressures. This then means that because the Moon is nearer the Earth than the Sun, its gravitational attraction on the Earth's atmosphere is considerably greater, and when the atmospheric bulge is facing the Moon, the atmosphere has actually expanded to form the bulge, so that the barometric pressures are not altered as the Earth spins on its axis within the Moon's orbit. The atmospheric depth alters but not its pressure. The lunar tide is there, but there is no way the tide can be seen. One can only infer, by logic, that it is there.

While the Moon is not above the horizon, the atmosphere will be slack on that part of the Earth away from the Moon, and under normal or natural pressures. The line on a barograph will remain fairly constant at that time, while the Earth continues to spin on its axis inside the Moon's orbit around the Earth. World-wide the atmosphere will rise and fall or expand and compress, but without much alteration in barometric pressures in any one place and whatever the pressures may be in that period.

5: *The inclined orbit effect*

Everyone knows that the Moon has phases, depending on the relative positions of the Sun, Moon and the Earth. The phases in order, are new, first, full and last. But what is not so well known, and repeating some of what has already been written, is that the Moon in its orbit around the Earth has a tilt, or it is inclined by about five degrees to the ecliptic (the plane of the Earth's orbit around the Sun). This tilt of the orbit has a movement referred to as a 'precession', which means that at the point where the Moon's orbit crosses over, and going either north or south of the Earth's ecliptic, is called the ascending or descending node and that crossing point will shift regularly to a new position on the orbital path of the Moon in each orbit of the Earth.

The shift is equal to a little over one degree an orbit.

When the crossing points of the Moon's nodes occur at each of the equinox positions on the Earth's equator, then the northern part of the inclined orbit of the Moon will be five degrees north of the ecliptic at the summer solstice position of the Northern Hemisphere and then inclined to be five degrees south of the ecliptic over the Southern Hemisphere. The five degrees is then added to the Earth's tilt of 23½ degrees so that the Moon's declination, or latitude on Earth, will be slightly above 28 degrees north and south of the Equator each month while the Moon is at the solstice position during the Moon's orbit of the Earth.

The effect on the Moon's declination during its orbit will be that as the Moon orbits the Earth, the new moon, which always occurs over the hemisphere enjoying summer, will rise higher in the sky than the Sun does at mid-summer, but the full moon will rise only about as high in the sky

Moon's path each Lunar Month from South to North of the Equator when the declination is 18⁰.

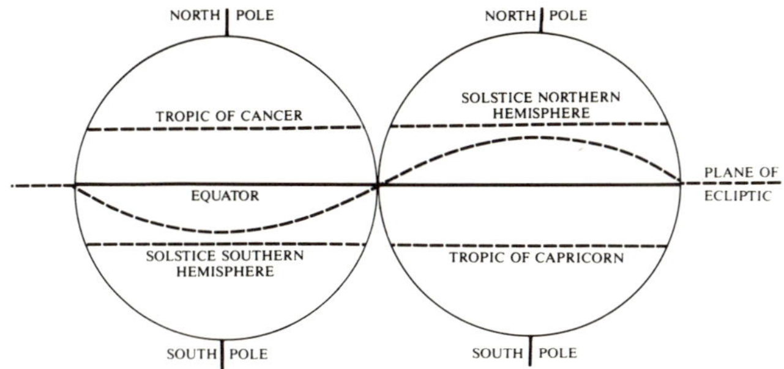

Moon's path each lunar month from south to north of the equator when the declination is 28⁰.

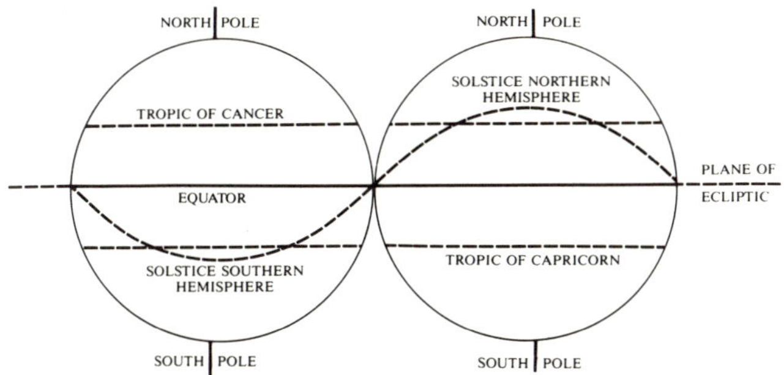

The effect on the moon's orbit after 180⁰ on its precession cycle of 18.6 years.

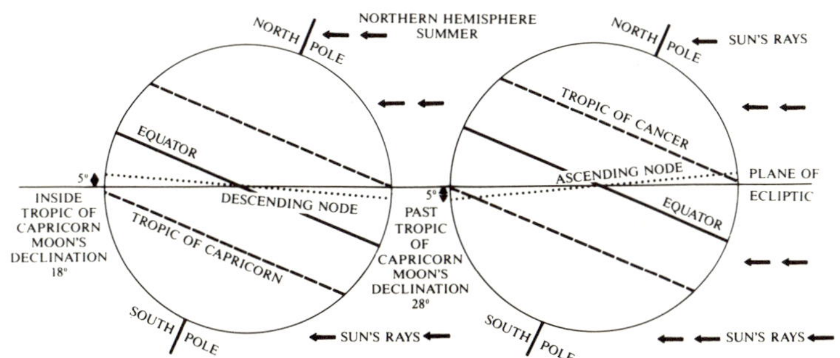

as a winter sun will do, when viewed in that same summer month. Each phase of the Moon, in its turn, as it moves into each solstice position during the year will be at that 28 degree declination.

A slow change, as the precession proceeds around the Moon's orbit, will see an opposite situation after about nine years when the north part of the Moon's inclination will change to be five degrees south of the ecliptic; also when the Moon is at the summer solstice position of the Northern Hemisphere and five degrees north of the ecliptic at the summer solstice position of the Southern Hemisphere.

What this means in practical terms is that the tilt of the Moon's inclination to the ecliptic is away from the tilt of Earth on its axis or in the opposite direction. So this time the five degrees is subtracted from the 23½ degree tilt of the Earth and the declination of the Moon (its latitude) rises to a little above 18 degrees north or south of the Equator, and the effect on the Moon will be, that the new moon in either summer and while at the solstice position, will not rise as high in the sky as the Sun does during the summer months, and the full moon in the same summer month will arc across the sky five degrees lower than a winter's sun. This will be for each phase of the Moon while it is at a solstice position at the appropriate time of the year for that phase, and again this will be for each lunar month.

The nodes' crossings will be at the same point as when the Moon was at its maximum declination — at the equinox position which is also the nodes' crossing of the ecliptic — but the ascending and descending nodes will be in opposite positions, 180 degrees apart. In between the maximum and minimum declination periods of the Moon, the nodes, as mentioned, move to a different position on the Moon's orbit until the declination of 28 degrees reduces to the 18 degree minimum.

In the first instance, where the Sun moves 23½ degrees north and south of the Equator, which is from the Tropic of Cancer to the Tropic of Capricorn, the Moon's declination as the movement is called, will be close to 29 degrees north and south of the Equator. After the nine years, the declination will reduce to something over 18 degrees north and south of the Equator, each lunar month, and once again the declination will slowly return to about 29 degrees. The full cycle is over a period of about 18 years.

Whatever the maximum declination of the Moon north or south of the Equator each lunar month may be in any particular year, between the 18 and 28 degrees, it will always take place at the same point, which equates with the summer or winter solstice position for the Sun and, as already mentioned, for each phase of the Moon in its turn during the year.

6: Can seasons be compared?

These astronomical details on the Moon's orbit are given to show the complexities associated with the Moon and its orbit around Earth, and to offer reasons why no obvious connection would have been seen between the weather and the phases of the Moon or on any part of the Moon's orbit. But as the phases of the Moon are the most visible feature we see when viewing our satellite, it is always the phases which most people refer to when making a comparison between weather and the Moon.

As weather of a type can be seasonal, then if a comparison has to be made from one year to the next, and to the nearest phase of the Moon as well, it must be remembered that there are about 13 lunar orbits in a year. To arrive at the nearest phase of the Moon after exactly 12 months, it is necessary to go forward another 19 days, but this is only one aspect where there is a change within the lunar orbit. The maximum declination date also will change so that the southernmost or northernmost point reached in the Moon's orbit will have moved forward by about 17 days. In fact, every position of importance on the orbit will have changed in relationship to the same date of a previous year.

Because there is always that attempt to compare or associate weather with the phases of the Moon, some demonstration on the futility of the exercise may suffice.

Using the year 1986 and the full moon as the example (all dates are GMT) and to the nearest date and to the nearest degree in:

		Moon's declination	North of the Equator
January	full	26	24 degrees
February	full	24	17 degrees
March	full	26	1 degree
			South of the Equator
April	full	24	9 degrees
May	full	24	24 degrees
June	full	22	28 degrees
July	full	21	26 degrees
August	full	20	15 degrees
September	full	18	5 degrees
			North of the Equator
October	full	18	11 degrees
November	full	16	19 degrees
December	full	16	27 degrees

The moon was at its maximum declination of about 28°43′ north and south of the Equator about September 1987-March 1988.

As a comparison, the minimum declination of 18°10′ which occurred in September 1978 now follows, but using the new moon instead of the full moon, a check with the previous table will show which hemisphere each moon occupied for the same month.

1978		Moon's declination	South of the Equator
January	new	9	17 degrees
February	new	7	13 degrees
March	new	9	3 degrees
			North of the Equator
April	new	7	3 degrees
May	new	7	12 degrees
June	new	6	18 degrees
July	new	5	18 degrees
August	new	4	14 degrees
September	new	2	9 degrees
			South of the Equator
October	new	2	1 degree
November	new	1	11 degrees
November	new	30	16 degrees
December	new	30	17 degrees

In 1986 the daily movement northwards as the Moon crossed the Equator was 6°9′, reducing of course as the Moon approaches the northernmost point in its declination.

In 1978 the daily movement at the equator was 4°12′ when going north.

As one will observe, full and new moons move in opposite directions for the same month of the different years. Also each of the Moon's phases change hemispheres with the season, the full moon being over the Southern Hemisphere during its winter and with the new moon there in the summer season.

Each quarter moon naturally moves in a similar manner, and each is at the maximum northern or southern declination at the time the Sun is in the equinox positions.

In the 1978 example, the Moon covers 36 degrees of latitude between the south and north declination and its daily advance northwards at the Equator is a leisurely four degrees or so.

In 1986 the Moon covers some 56 degrees of latitude in the same period of time, in about 14 days, and its daily advance at the Equator is a quicker six degrees or so.

These figures will show the futility of comparing random years with one another in seeking an answer to the question of whether or not the Moon does influence the weather.

The importance of the Moon at perigee or at apogee must not be lost sight of either. Some of the perigee/apogee cycle is covered on pages 14-16.

Perigee/apogee effect

But perigee like the Moon phases, also advances around the orbit in a cycle lasting 8.85 years, which does follow the Moon phases in direction, but of course not at the same pace.

Each Moon phase throughout a year will be associated with perigee in its turn, as the Moon orbits Earth. It follows that whichever phase of the Moon is at perigee, that phase will be at its closest approach to the Earth at that time, and this close approach will have more influence on the atmosphere at that time than the Moon would have if it had been at apogee (the most distant point).

Also, the perigee moon will spend about half of its 8.85 year cycle over each hemisphere, and of that half, much of it will be at or close to the maximum declination point. So the stronger attraction from a perigee moon will pull more of the atmosphere towards that hemisphere and away from the opposite hemisphere. This attraction will be more pronounced while the perigee moon is at either of the maximum declination points north or south of the Equator.

If the Moon's phase is new and at perigee, a summer moon over whichever hemisphere, then the combined Sun and perigee moon attraction can leave the opposite hemisphere with a very low atmospheric tide, and in many instances this is when the very bad storms occur over that hemisphere from the low tide; especially when the Moon is at its maximum declination of 28 degrees because it is at this time that the Moon is beyond the latitude of the tropics by five degrees.

Where the Moon will rise

While the Moon is at the 28 degree declination, the differences in the Moon's arc in the sky on rising and crossing from east to west is best observed during the time of either the summer or winter solstice. It is at this time that the new and full moons are at their maximum declination points, either north or south of the Equator. In the Southern Hemisphere when viewing the new moon of summer, it will rise well to the south-east, climb high in the sky, and set well to the south-west in the same manner as the Sun will do.

The full moon, in the same month, will rise to the north-east and rise about as high as the winter sun, then set in the north-west.

In the Northern Hemisphere the directions are reversed, for the new moon in the summer, rising well to the north-east and rising almost perpendicular to its zenith, then setting well to the north-west, and for the full moon, rising in the south-east and setting in the south-west.

These rising and setting points will slowly move towards the Equator as the Moon's declination declines to the 18 degree declination after about nine years.

Did astronomers fail meteorologists?

Although meteorologists and astronomers say plenty in their own fields of interest, neither section has raised perhaps the most important factor in atmospheric movement.

The mention of astronomers in this area is necessary because gravity and gravitational attraction is supposed to be in their area of expert knowledge.

Astronomers will explain what is happening out in space. They will tell us why clouds of dust act in a certain manner, where the clouds have come from and where they are going to, but at their own back door these astronomers did not know what was happening to, or in, our own atmosphere and why it behaves in the manner that it does.

They will agree that gravitational attraction will affect clouds of dust out in the universe untold light years away, but they do not recognise that gravitational attraction from our next door neighbour, the Moon, and our sun must have an effect on our atmosphere. It seems that the astronomers also believe the barometer.

The Earth circles the Sun on a stable plane. Earth moves around that orbit with an axis tilt of 23½ degrees or 66½ degrees to the plane of the ecliptic with the tilt always facing in the same direction, and ignoring for the moment the possible long term effect of the Milankovitch Model (an explanation of this model will be covered later).

The Moon orbits Earth on the plane of the ecliptic, but with a tilt or inclination of five degrees which alters the position of its tilt on its orbit as has been explained previously. Within the Moon's orbit, the Earth spins once each day.

If it is agreed that the Sun does have a gravitational attraction on the atmosphere, then that attraction will be directly in a line between the Earth and the Sun and that attraction will be all encompassing, not just horizontal to the plane of the Earth's orbit around the Sun, nor vertical or angled.

As mentioned, the Earth is tilted or angled to the plane of the Earth's orbit around the Sun and this means that for much of the year, the Earth's daily rotation on its axis (and this naturally must include the atmosphere) spins on an angle through this Sun-Earth orbital plane, the ecliptic. This twisting action must cause some distortion within the atmosphere.

Because the Earth's tilt is always facing the same direction, the tilt will have less effect on the atmosphere for much of the summer solstice as that segment of the Earth facing the Sun will be rotating on its axis through the same orbital plane of the 'Sun-Earth' or along the latitude lines on Earth 23½ degrees north or south of the Equator. In the Northern Hemisphere this will apply to most of the months of May, June and July; in the Southern Hemisphere, November, December and January. This action will apply to the winter solstice as well, but the months will be reversed.

Astronomers missed this effect

During months near the equinoxes, the twisting action through the plane of the ecliptic disturbs the atmosphere in varying amounts.

In the temperate zones, and probably in the higher latitudes too, the equinox period is the time for what is commonly called the 'Equinoxial Gales'.

North and south of the Equator and within the tropics, this is also the season for the destructive storms which are known variously as cyclones, typhoons or hurricanes.

While the Earth is revolving on its axis each day, that area occupied by the trade winds north and south of the Equator will move through the plane of the Earth's orbit around the Sun (the ecliptic) twice a day — once on the way down and again on the way upwards. The northern Tropic of Cancer will drop in a vertical distance 47 degrees north of that plane, down to the ecliptic, which is the horizontal line of that plane, and the Tropic of Capricorn will move from that ecliptic to the same distance south of that plane. So it can be seen that there must be a considerable amount of distortion applied, by gravitational attraction, to the atmosphere.

Is all this meteorological activity at this time of the year a coincidence? It would seem that it is not, and that gravitational attraction must be the most important reason for this activity.

Now if the Sun were the only gravitational influence on the atmosphere, then some regularity in the weather of the world might be expected. Everything would be orderly, following an easily predictable pattern. Astronomically, each change between the Sun and the Earth throughout the year would be the same, year in and year out. But because we know that this is not the position relating to the world's weather, some other gravitational influence must also be at work. As the Moon is the only other close celestial body to the Earth, it must follow logically that it is from this source that we receive the greatest distortions to the world's weather.

Put briefly, the Moon is the disturber and the distributor of the world's weather systems, irrespective of what other influences may be at work.

The 23½ degree tilt of the Earth offers some intriguing thoughts when it is related to the relative position of the Sun and Moon during the various seasons of the year. For instance, when the Sun is positioned at the vernal equinox (March 21), the Earth's spin through the plane of its orbit around the Sun is from north to south. This means the Equator is 23½ degrees north of the Earth's plane (the ecliptic) and passes through the Earth's plane to 23½ degrees south of the Earth's plane, which should mean that the atmosphere could develop a north-west drift from the gravitational attraction of the Sun.

Now, with the Moon phase at full, which places the Moon over the opposite equinox position where the Earth's spin is upwards from south to north through the plane of the Earth's orbit around the Sun, this movement may give the atmosphere a drift to the south-east on that side of the Earth.

With the Sun near the same position but the Moon at the last quarter, the Moon would be at the summer solstice position for the Southern Hemisphere. The drift to the north-west from the Sun's influence would remain the same, but the Moon's attraction or drag on the atmosphere will follow the latitude lines, or a drift to the west by the atmosphere. This means the west drift from the Moon's influence will change to north-west from the Sun's influence after 270 degrees of the orbit are covered; the last quarter moon rising before the Sun at about midnight.

The gravitational attraction of the Sun and Moon would then be contained within approximately 90 degrees of arc, while the other 270 degrees would be relatively free from gravitational interference. The 90 degrees of arc takes in the angle between the zenith of both the Moon and the Sun. Obviously there would be some gravitational influence both before and after the zeniths, but it is felt the greatest attraction would be within the angle of 90 degrees, and less so outside that angle. With a new moon at the same equinox with the Sun, the combined gravitational attraction on the atmosphere could have a marked effect at that time, especially the possibility for stronger winds or equinoxial gales.

During the night hours the Earth will move into an area without gravitational attraction and a low atmospheric tide, which may explain why the nights at this time are usually cool and with the winds less strong or even calm.

With the Sun still at the March equinox, a first quarter moon will be over the Northern Hemisphere at the solstice position and the 90 degree segment of the arc of the Sun and Moon will take in the angle from the equinox to the northern solstice position. In this instance the drift to the north-west will change to west because the Sun will rise before the Moon.

The main points of interest are the varying positions of the 90 degrees of arc. It does not matter whether the Sun is at the solstice positions and the quarter moons respectively at either of the equinoxes or the other way round with the Sun at either equinox and the Moon at one or the other of the solstice positions. It is at these times the most torque is applied to the atmosphere, and when related to the inclination of the Moon when the node crossings are in the north and south solstice positions, it could offer sound reasons why there seems to be so much mixed up weather during that period referred to as an 'oscillation'. The subject of the oscillations will be covered in more detail later.

7: *Sunspots*

Most meteorologists and climatologists when discussing or writing on their respective subjects, mention sunspot activity and its effect on the weather or climate.

Sunspots are large areas on the Sun's surface with temperatures that are apparently cooler than the rest of the Sun's surface. The sunspots increase and decline in numbers over a cycle which is generally recognised as averaging about 11 years.

When sunspots peak, astronomers and meteorologists are of the opinion that they affect the Earth's weather and some very interesting observations have been made, mostly in the area that they cause droughts in many parts of the world.

The sunspots create strong solar winds and these act on the Earth's magnetic field and produce the Aurora Australis and Aurora Borealis. Some scientists have connected the aurora displays with increased depression activity over the high latitude areas.

The sunspots also have been associated with restricted tree growth and it has been claimed that the growth rings in trees, which have been felled for research, were found to be thinner after a period of maximum sunspot activity, apparently because of the associated droughts claimed to occur with maximum sunspot activity.

In the USA, tornadoes have been associated with sunspots and a cycle has been observed where tornadoes had occurred in one state and over a cycle of about 45 years progressively followed a circular path through other states and returned to the original starting point. Why such a cycle should be in existence is difficult to follow.

A lunar effect would be more logical but without knowing the exact dates and where the tornadoes took place, even that suggestion could be doubtful. Someone may have taken just the tornadoes which did follow this cycle and ignored every other tornado which took place in other areas during the period of the cycle. It would be most unusual if every tornado was on the cyclic path and nowhere else during the 45 years.

The sunspots commence their existence near the Sun's poles and gradually move to the equatorial region where they diminish. The cycle recommences again in about 11 years but the magnetic charge associated with the sunspots changes polarity, so that the sunspots will be charged negatively or positively alternately for each successive cycle.

How this alteration in polarity can affect the weather on Earth, no one can say. About the only suggestion that could be offered is that the cloud formation on Earth has a polarity which is, at times, repelled by the solar wind, thereby in some manner discouraging rain, or possibly encouraging rain. Who knows? For if there was a magnetic attraction by a cloud formation to any solar energy entering the atmosphere, it could be argued that it may encourage rain, or as stated it may not. Some sensible research could be developed from this idea and may offer a more positive answer to this question.

Although the sunspot cycle is said to be about 11 years, the time span can be variable and it may be as low as nine years or up to 15 years and sometimes longer.

In the past there have been long periods when there has been little or no sunspot activity and these periods have been connected to long spells of cooler global temperatures with increases in glacial advancement, while maximum sunspot activity is said to cause glaciers to retreat.

It has been mentioned in another chapter that the Moon has a precession cycle changing the declination from a maximum of over 28 degrees and then reducing to about 18 degrees in nine years before returning to 28 degrees, the total cycle is over a period of 18 years.

The first year when the Moon's declination was at 28 degrees which could be connected to weather records held by me, was the year 1914. The previous cycle would have commenced about 1896, which is just outside my oldest records.

The records as listed, and taking in the two years close to the 18 year cycle, are for Hamilton in New Zealand's North Island. There has been no attempt to gain information on any other area in New Zealand, and it may be that the following comments are not completely applicable to, say, the South Island, but the facts and figures may do so in part anyway.

A comparison below is between the rainfall statistics for when the Moon's declination is at 28 degrees and when the Moon is at the 18 degree declination.

The average rainfall for Hamilton is about 1193mm or near 47 inches.

Year	Total Rainfall	No. of days with recorded rain
1914	731mm or 28.78 inches	176
1931	1068mm or 42.06 inches	144
1932	842mm or 33.16 inches	134
1949	1014mm or 39.93 inches	153
1950	1005mm or 39.58 inches	132
1968	1281mm or 50.46 inches	186
1969	1120mm or 44.09 inches	157
	Totals: 7061mm (average of 1008.71mm)	1082 days over 7 periods, (average 154.57 days)

Strangely the sunspots were at their maximum in 1968 when, if theory is correct, the weather should be drier, but all the other years show lower than average rainfall totals.

The yearly averages could have been broken down into summer and winter figures. Had this been done, the summer rainfall figures, including days with recorded rain, would show as being extremely low, and this in part was the same for 1968.

The 28.78 inches of rain recorded in 1914 was an all-time low, and no other year, in the records held by me, has recorded less rain.

The doubling of the years indicates that the maximum declination of 28 degrees covered both years.

As a comparison, the 18 degree minimum declination of the Moon north and south of the Equator follows:

Year	Total Rainfall	No. of days with recorded rain
1903	1511mm or 59.5 inches	Not known
1904	1651mm or 65 inches	Not known
1921	1140mm or 44.9 inches	186
1922	1200mm or 47.26 inches	186
1940	1008mm or 39.70 inches	148
1941	1306mm or 51.43 inches	154
1959	1218mm or 47.96 inches	176
1960	1306mm or 51.44 inches	177
1977	1191mm or 47.24 inches	195
1978	1085mm or 42.66 inches	158
1979	1247mm or 47.64 inches	192
	Totals: 8369mm from 1921/77 (average 1195.57mm)	1222 days for 7 periods (average 174.57 days)

In comparing the two lists, it becomes obvious that in general terms the low declination offers higher rainfall figures and higher number of days with recordable rain. The number of days with rain for 1903/04 was not in my records.

The question which needs asking is: How does sunspot activity fit in with these figures? Sunspots were at their maximum in 1976-79 years, yet the rainfall locally was up to the average except 1978, and the days with recordable rain were also higher than usual. No averages are mentioned for days with recordable rain, but most years here are about 175-180 days with recordable rain.

8: *Pointers to weather predictions*

People recording meteorological information for the New Zealand Meteorological Service are supplied with charts on which are entered the amount of rain recorded on each day rain occurs.

The charts are for 12 months and the months January to December are across the page and the days are vertically down the page with lines forming a block for each day each month.

Once the charts are filled in progressively, one can almost say when rain may be likely during the following month, because the blocks with the filled in rainfall figures frequently occur in the same area across the chart. One cannot guarantee that it will rain, but the chances that it may do so will be enhanced if the preceding months had rain at that time.

Because rain is recorded, at times in the same part of the month each successive month, there seems to be an implication that a lunar influence could be the reason for the rain arriving during that part of the month each month.

If a lunar influence is rejected, then what else!

There is no other regular earthly movement to account for such regularity. The Earth rotates once a day, which adds up to a month, but this month could be varied to any number of days by government legislation. There is no valid reason that a month must contain 30 or 31 days. The number of days in a month for a year only needs to add up to 365 days in total.

For instance, 14 months of 26 days = 364 days, so one month of 27 days will give the 365 days for the year.

However, the dates for each phase of the Moon, when related to the calendar months presently in use, will show only a slow change from one month to the next for the same phase, and with only an advancement of a day in most instances, so this slow change with Moon phases dates, month by month, does indicate a lunar connection with rainfall when checked against a meteorological rainfall chart.

This connection between the lunar phases and rain may be denied but where can one go to receive a better answer than this one?

Time does not permit experimenting with other aspects of the weather records, but if the temperatures and barometric pressures were to be recorded on a chart in the same manner as the rainfall figures, would a pattern similar to the rainfall chart show through with temperatures and barometric pressures indicating a lunar effect?

Blocking anticyclones — their causes

If there is an area within meteorology on which some clarification is needed, it is on that expression called the 'blocking anticyclone'.

In practical terms it means a stationary anticyclone or one which is slow moving.

An anticyclone so situated probably is 'blocking' the progress of the following weather systems, but are not the following systems blocking systems even further west, or perhaps the 'blocking anticyclone' has been blocked by other systems further to the east.

The use of the term means that there is no real understanding on what is happening to the atmosphere or why it is happening.

All weather systems, be they anticyclonic or depressions, will become stationary occasionally — even troughs of low pressure at times will be reluctant to move away — or these conditions will remain much the same with fronts crossing an area regularly for a period.

If the Moon is accepted as being the controller of the atmospheric movement, then it must be some action by the Moon which is slowing down the weather systems.

When the Moon is moving either north to south or vice versa through the Equator, the daily movement in a vertical line — the declination is variable through the latitudes and changing slowly from about 6°11′ at the Equator to about 0°.31′ at the southern or northern most declination point if the Moon is at say, the maximum declination of 28 degrees.

The horizontal line of the Moon's orbit at this southern or northern position almost follows the latitude lines for several days before angling say, northwards from the southern declination and through the lines of latitude. The Moon then goes through the same movement at the northern declination point.

During the 18 degree minimum declination, the daily movement of the Moon at the Equator is 4º18′ and at the southern declination the movement is 0º.11′, much the same as for the 28 degree declination, but in this instance the Moon advances northwards on a lesser angle through the latitudes as it orbits the Earth.

The Moon's movement through the latitudes can be likened to a right-hand thread of a nut slowly being turned upwards, then downwards on a thread; except to get the correct action the downward movement would equate with a left-hand thread as seen from the opposite side of the Earth.

Coarser threads on a bolt, such as a Whitworth thread would equate to a 28 degree rise and fall, while a finer thread such as SAE would represent the lesser lift of an 18 degree declination.

It is during this southern or northern declination section of the Moon's orbit, which is also the solstice position, that the movement of the atmosphere, in most instances, is restricted and the atmosphere will become slow moving and any anticyclones or any other weather system may stand still, or almost so, thus giving rise to the meteorologists' statement that there is a 'blocking' anticyclone.

When the Moon phase is approaching new or full, a slowing down of the atmosphere can take place at whatever section of the orbit the full or new phase occurs and this slowing down may be away from the solstice position. But as the new or full moon approaches the solstice position, the weather systems, be they anticyclonic or low pressure, will slow down under the influence of the new or full moon anyway. The Moon will then advance into its maximum declination point which is also the solstice section of the Earth's orbit around the Sun, and may be either the summer or winter solstices.

The new moon is always at its maximum declination at the summer solstices for either hemisphere and the full moon for either of the winter solstices. At the maximum declination point the weather systems, usually, will remain stationary for several days before resuming their easterly movement.

A new moon position will advance about 30 degrees forward in its orbit from one month to the next month on its orbit around the Earth. Over, say, the Southern Hemisphere, a new moon may be about 30 degrees short of the southern declination point in November, so the slowing down of the atmosphere or weather systems may commence a couple of days before the new moon, then another two days to the southern declination, followed by two to three days before the weather systems begin to move after passing the influence of the southern declination.

In all, six or seven days will be involved when the weather systems may not move very much, be they anticyclones or any other weather system.

In December the slow down period will be less because the new moon will be at the solstice position and the slow down periods will merge, but the slowing down of the atmosphere will be likely to increase again in January as the new moon arrives at that point 30 degrees past the southern declination or solstice position when that slow down point will be added to that of the new moon.

Although the figure of 30 degrees before the declination point has been mentioned, astronomically the new moon could be, say, 40 degrees short of that point in November and still cause the atmosphere to slow down, and as the Moon advances in its orbit at about 14 degrees a day towards the southern declination, this would involve about two days before new moon, three days to the southern declination, and another three days to advance past that point. So eight or nine days could elapse before a significant movement in the weather systems would be noticed.

The same conditions may occur again in January if the circumstances of November are projected to that month.

If the weather systems happened to be anticyclonic, dry weather would follow.

Similar conditions could apply over the Southern Hemisphere while the Moon was at full during the winter months, and coupled with a new moon slow down in the same month, a significant part of the month could be affected and there would be no guarantee the rest of a month would be much different.

From this explanation it is easy to understand how and why drought conditions may commence.

For the Northern Hemisphere read: May, June and July months for summer.

During the middle winter months of either hemispheres, the situation could be similar, with either an extended dry or wet spell depending on the weather systems operating at that time. However, because the new moon is over the opposite hemisphere to that experiencing winter, and because both the Sun and the Moon are together over the summer hemisphere, the atmospheric tide will be low over the winter hemisphere, so the weather, more than likely will be unsettled more often than not at that time.

Wherever the new or full moons take place on their orbits outside of being at the solstice positions, the weather systems will still have a tendency to slow down for a day or two, and the slow down will also occur while the Moon, at whatever phase, passes through the solstice or maximum declination part of the orbit although the slow down is not usually lengthy.

The Moon's effects on systems

If further evidence is required on the Moon's influence on the weather systems, it will be provided by taking note of which way, say anticyclones, drift.

Over the Southern Hemisphere when the Moon is at or near the southern declination, and at whatever phase other than new or full, the anticyclones will drift slowly along the latitude lines.

Should the Moon be moving upwards towards the Equator, the anticyclones will begin to drift slightly north-east. When the Moon moves downwards from the Northern Hemisphere after crossing the Equator on its southward journey, an anticyclone will be inclined to drift towards the south-east.

Depressions may be affected also, but to a lesser extent. They will drift south-east as the Moon moves southward, but on the Moon moving northwards, the depressions are inclined to drift more east and only occasionally towards the north-east.

Once the Moon has crossed the Equator, the drift of the anticyclones over the Southern Hemisphere seems to be less predictable, but usually they will drift to the east.

Over the Northern Hemisphere, and with the Moon over that hemisphere, the drift of the weather systems is to the east the same as for the Southern Hemisphere, but change north to south and south to north in each instance, whether in association to the east movement or directly north and south.

Depressions at lower latitudes, say near the tropics, seldom drift away to the south-east over the Southern Hemisphere or north-east over the Northern Hemisphere while the Moon is moving towards the Equator in either hemisphere, but more often than not they will drift directly east.

The movement of tropical cyclones is also governed by the Moon, and although the Sun provides the heat for their formation, the phase of the Moon will dictate how much heat will be provided to get these massive and destructive storm systems rotating. More information and facts on cyclone/hurricane movements will be written into another chapter.

9: El Nino

In parts of the world, from time to time, massive atmospheric aberrations occur, stretching for thousands of miles. One which dominated the weather over the South Pacific in 1983 and lasted for about 12 months, is called El Nino after a local current along coastal Peru and Equador. It usually occurs near Christmas. The title means 'the child' although other interpretations have been offered.

The waters of El Nino gradually warm up and spread westwards over the Pacific. Meteorologists could not agree on why the waters became warmer. Some were of the opinion that a shift in the current allowed the warm waters to move into the area, others thought that the atmosphere was the reason by transferring heat to the sea. Actually heat can only be transferred from or by the way of the atmosphere to the sea, but less so in reverse. If the atmosphere remains cold over a warm sea, the sea will cool slowly and if the atmosphere remains warm over a cold sea the sea will slowly become warmer.

During the commencement of El Nino the barometric pressures will begin to alter from the normal pattern and oscillate over the Indian Ocean and across to the central and the east Pacific. The changing pressure patterns cause what is referred to as 'The Southern Oscillation' and its effects are far reaching.

Cyclones in the central Pacific were more frequent and damaging than for many years. Changing wind patterns brought a severe drought to Australia and South Africa, with storms and flooding along western sea coasts of both North and South America.

Sea temperatures were considerably below normal along the east coasts of the North Island of New Zealand during the summer months, causing algae bloom to accumulate because winds were persistently westerly, leaving the east coasts sheltered. This in turn created problems for sea bottom feeding animals which died when the bloom covered the sea floor and their food supply.

Meteorologists and climatologists had a field day, which lasted as long as the 'Southern Oscillation' and articles were still being written on the subject three or four years or more later. Every unusual weather pattern since then has now been related to El Nino whether it be a sudden snow storm, lightning, and even, in New Zealand, if south-west winds were more frequent than usual. It was stated, at one stage, that high pressures over Australia caused these winds as if the winds were from that direction continuously when south-west winds are a normal expectation every now and again.

Taken to its logical conclusion it must mean that El Nino is always with us so that it can be used to offer an explanation for all otherwise unexplainable phenomena.

More was said and written on why the weather patterns existed than for any other single event in meteorological history, but none could say what brought on the oscillation.

Many theories were brought forward on what may start the oscillation in the first instance, but until the barometric pressure anomalies commenced the meteorologists are unable to predict when the pressure changes will take place.

Although the pattern for the 'Southern Oscillation' has been well known for about 50 years, the term has not been used commonly before this last oscillation, and apparently, to my knowledge, newspapers had never made use of the term previously either.

I have diligently kept anything that refers to weather since about 1962, and on going through my files, only one other period has mentioned crazy mixed up weather (as distinct from the other comments brought together by myself later in this section). There was comment on an unexplained shift in the Humboldt Current on the west coast of South America, nothing on El Nino or the 'Southern Oscillation'. So, as far as the general public was concerned, the 'Southern Oscillation' was a new phenomenon occurring during the years 1982/83.

At this time, 1982/83, the crazy mixed up weather included major storms on the west coasts of the United States, Australia suffered its worst drought in two centuries, French Polynesia had its first cyclone in 75 years and five more in as many months. Peru had its highest rainfall in 450 years and records were shattered in Ecuador. China had floods in the north and droughts in the south. Many other countries also had climatic problems at this time.

The 'Southern Oscillation' was said to have been responsible for 1300 to 1500 deaths world-wide and caused an estimated $NZ3.7 to $14.8 billion dollars worth of damage.

Recently meteorologists have been studying a connection between a 40 to 50 day oscillation and El Nino, both of which have rising air currents known as convections which are reported across the Pacific in regular cycles. It is thought there may be a connection between this event and the 'Southern Oscillation'. There may be a connection in a very minor way but for it to develop into a world-wide destructive force appears unlikely except in the manner which will be explained in this chapter.

These minor oscillations occurring from time to time could be the excuse or reason for some meteorologists to keep on referring to El Nino on every occasion some unseasonable weather changes take place.

What happened in 1974/75?

In 1974/75 there were repeated references to 'strangest weather of the century', 'crazy mixed up weather', 'warmer weather in Auckland over 29 months caused by shift in the Humboldt Current', 'fish leave Peru stranded', 'Peru's fishing industry in parlous plight', 'weather has experts confounded'.

All this seems to add up to a 'Southern Oscillation' or 'El Nino', but on those occasions it was the Humboldt Current which was highlighted in the newspaper articles.

Was 1965/66 similar?

Going further into history, this time 1965/66, my files show that world-wide there were severe storms, cyclones and droughts, but there were no articles collating these world-wide disasters, no mention of a 'southern' or 'northern' oscillation, for many of the storms affected the Northern Hemisphere as much as the Southern Hemisphere.

In New Zealand some flooding, mostly in the north of the North Island in 1965, snow to low levels in June and July — the winter was considered severe — and a summer that wasn't for most of the country, it being cold and wet up to December, especially in the North Island. Gales were fairly frequent.

The winter of 1966 was still wet in many places and with snow killing lambs in August.

In Australia, bush fires in March were serious in New South Wales. The area was in the grip of a drought and farmers walked off their farms. In

Victoria, a similar story. The drought broke in Brisbane late in April but was still severe in June; then followed gales, some snow and freezing temperatures with a severe winter, but the drought was not really broken, only eased.

In August 1965 heavy rain fell in Central Australia.

By September temperatures began to soar again in New South Wales while in November, South Australia experienced gales, hail, snow and also torrential rain.

Gales affected New South Wales later in November.

The drought returned to Queensland, but New South Wales received heavy rain late in November, and Central Australia became an inland sea. A few days before it was a dust bowl.

It was not until April that storms finally broke the drought in New South Wales; eight inches of rain fell in 24 hours, and caused widespread damage.

In the Pacific, Fiji suffered from a cyclone in February 1965 and Samoa, February 1966.

Tonga suffered drought conditions and received its first rain in two months in mid-December 1965.

Fiji also was affected but rain did not arrive until mid-January 1966. A storm in December affected only the south-west.

South Africa was in the grip of a drought in March 1965 as was Kenya. By January 1966 it was referred to as the worst drought this century. Some rain in Kenya during January eased the drought, but it continued over the rest of southern Africa.

Along the west coast of South America a storm in August of 1965 — the worst this century — brought widespread rain, killing many people in Chile, and affecting areas from Coquimbo to Puerto Montt. Mention was made of a giant flood on a 1000 mile front, while another big new storm in the Pacific was heading for land.

The sun did not shine for 15 days and in September a cyclone swept over the area.

In December heavy rain caused a landslide which killed at least 60 people in Peru.

Over the Northern Hemisphere a similar story may be written. Over England and Europe in March 1965 blizzards with 10 foot snowdrifts. In April spring turned to winter with snow, rain and bitter winds. It was Easter and comments said the worst in memory.

In May 1965 a mini heat wave, London 82°F, Paris 83°F. In north Italy there were drought conditions.

In July 1965, southern England, monsoon style thunderstorms swamped the area. One inch of rain in London's lunch hour.

In September 1965 an Indian summer in England.

Early in November 1965 gales hit Britain, north France and Scandanavia.

In mid-November 1965 wintry blast covering vast areas with snow. At the end of November, arctic conditions gripped Britain with gales across France.

Early in December floods, snowstorms, sea dramas, head the news for Britain and Europe. Late in December 1965, 90mph gales and heavy seas lash Britain, more snow, and floods in Wales.

In January 1966, Britain was in the grip of a big freeze after blizzards swept the south England, Wales and the west country, and also west Europe.

By mid January temperatures 'soared' to 3°F. The sea froze for 45 yards off the south coasts of England. In Paris a temperature of 8.6°F was recorded, the lowest since records began in 1873.

In August 1966, 100mph rain squall flooded towns in Wales. In the Midlands the worst flood for 30 years.

In Yugoslavia in May 1965, widespread floods. In Hungary in June 1965 the Danube in flood and vast stretches of Hungarian Plains covered, the worst for 11 years.

In Austria flood waters of the Danube and other rivers cover farmlands. While in Rome people die in a heat wave, with temperatures above 95°F.

By early September violent storms lashed central and north Italy and then Sicily. More than 40 people were killed. Late in September more floods hit Italy and Austria and the Tiber River overflowed. In France the grape harvest was threatened by a stormy wet summer.

In November 1965 snowstorms raged across Europe, mostly affecting the Balkans and North Italy.

Early in November 1966 a third of Italy was under flood waters and mud with over 100 lives lost and thousands hurt.

In May 1966 flood waters engulfed suburbs in Archangel, a northern Russian port.

In October 1966 torrential rain with floods left 50 dead and 7000 homeless in Algeria. The rain fell virtually unabated for six days.

In November 1965 the temperatures in Scandinavia were the coldest for a century.

In February 1966 Sweden had the worst winter for 25 years. Coastguards said conditions were the worst in living memory. Ships could not batter through ice walls 40 feet high.

Across the Atlantic in mid-June 1965, flood waters were 15 feet deep in the Texas town Sanderton, drowning 14 people and at the same time in Colorado, tornadoes and torrential rain caused flooding with 20 feet of water over the highway between Denver and Colorado Springs.

Late in January 1966 the worst winter storm since the 30s buried the eastern states of USA under more than a foot of snow. This was followed by heavy rain and melting snow, affecting areas from Texas to Minnesota on the Canadian border. Winds were up to 100mph and serious flooding occurred.

Early February 1966, a cold wave gripped North America with a death roll of 155, more snow was forecast. The town of Oswego in New York State had 90 inches of snow over six days.

By July 1966 a heat wave baked New York in temperatures near or above 100 degrees.

In November 1965 record-breaking rains swamped southern California in a sea of mud slides and flooded streets.

Across to East Asia: in South Korea in mid-July 1965 floods were caused by the worst rains in decades — 2600 houses were washed away and six people drowned. This after a long drought reported to be the worst in 60 years.

Near the end of July 1965, and still in South Korea, people died when persistent rain continued for more than a week with flooding; the number of dead was put at 47 with 15 missing and 37,800 homeless.

At the end of August 1966 on the island of Lombok east of Bali, about 50,000 people died from famine after a long drought.

In October 1965 cyclonic storms in central Burma lashed the area for four hours, killing at least 100 people.

In May 1965, 15,000 killed by a cyclone in Bangladesh and in July heavy flooding from monsoon rains. By October 1965 in Pakistan five million people were facing a severe famine.

In July 1966 in Bombay, India, heavy rain fell and it was hoped it would ease a drought there caused by the apparent failure of the monsoon. By August 30,000 villagers were evacuated from 45 villages around Delhi. People waded through waist deep water to reach evacuee camps. Two days of heavy rain swelled the muddy waters of the River Jumna to nearly three-quarters of a mile wide.

In September 1966, 1500 square miles of Bangladesh flooded after torrential monsoon rains.

In November 1966 India was in the grip of a severe drought after abnormal failure of rains from mid-August onwards.

During 1965/66 Japan suffered from several typhoons; May 1965, August 1965, September 1965. In September 1966 there were two. Taipei had one in June 1965. Miami had three, two in September 1965 and one in early October 1966, and Mexico one in late September 1966. The late September hurricane in Mexico could be the same one mentioned for Miami in early October.

A cyclone went up the east coast of USSR in November 1966.

Bangladesh had two cyclones, one in December 1965 and another in October 1966.

This list of meteorological phenomena is from newspaper cuttings in my files and it does not follow that the list is complete, but it probably highlighted those extreme atmospheric changes which occurred from time to time.

As mentioned earlier, there were no articles bringing all the problem storms, droughts and snowstorms together and questioning possible causes. Perhaps meteorology was still searching for answers. But my question is: was the weather that was experienced world-wide during 1965/66 an oscillation, and if it was an oscillation was it a southern or a northern oscillation?

The more recent one referred to as the 'Southern Oscillation' in 1983, seemed to affect the Southern Hemisphere more than the north, but the abnormal weather of 1965/66 had a profound effect on the Northern Hemisphere and a lesser effect over the south, although the problems created in some countries were still traumatic enough for those caught in those climatic problems.

Was the Moon responsible for El Nino?

The purpose of this book is to show a connection between the weather and the Moon's influence on the weather, and as these oscillations show the more disturbing elements in weather distribution, the next necessary question is how, or possibly why.

As has been stated earlier, the Moon orbits around the Earth close to the plane of the ecliptic (the Earth's orbit around the Sun) with an inclination of five degrees, in other words about half of the Moon's orbit is above the ecliptic by five degrees during a lunar month and half below the ecliptic by five degrees also. So as the Moon orbits the Earth each lunar month it crosses the ecliptic twice, once on the way up and again on the way down.

These crossing points are referred to as 'nodes'. The crossing point where the Moon is moving southwards is the descending node and on the way north it is the ascending node.

As mentioned previously, the Moon's five degree tilt to the ecliptic is called its inclination which is added to or subtracted from the 23½ degree tilt of the Earth's axis. Also, when the inclination and Earth's tilt are added to one another or subtracted and, as mentioned elsewhere, the two astronomical factors combined are related to positions on Earth, it then becomes the Moon's declination or latitude (the total vertical distance covered by the Moon north or south of the Equator). This distance varies from about 18 degrees north or south of the Equator and increases up to about 28 degrees over nine years approximately. So that over a period of about 14 days, the Moon when at its maximum declination, will go from 28 degrees south of the Equator to 28 degrees north, and this is represented in the 23½ degree tilt of the Earth on its axis plus the five degree tilt of the Moon's inclination to the ecliptic (the plane of the Earth's orbit around the Sun).

At this time the Moon's nodes (the point where the Moon crosses in either direction over the ecliptic) are at that point which equates with the equinoxes.

During nine years the Moon's tilt will slowly precess, so that the nodes will move around the Moon's orbit and eventually change places with each other at the equinox points. This precession will be over 180 degrees of the Moon's orbit. At this time the five degree tilt of the inclination to the ecliptic will be subtracted from the 23½ degree tilt of the Earth's axis so that the Moon will now move vertically from about 18 degrees south of

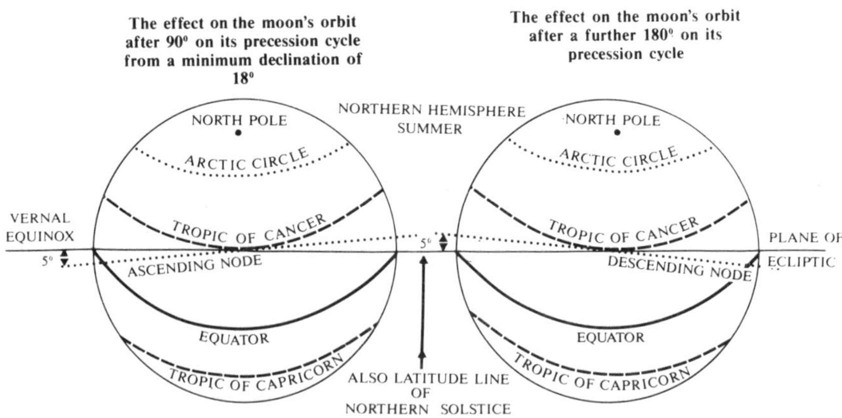

the Equator to about 18 degrees north of Equator, also over about 14 days. There are variables in the number of days taken for the Moon to move from south to north, and vice versa, it depends on where perigee occurs on the orbit when the Moon then will orbit at a faster rate and so reduce the time it will take to move from one position on its orbit to another, remembering that as the Moon circles the Earth it is also rising or lowering itself on the imaginary lines of latitude on the Earth and the faster orbital speed at perigee means a faster movement by the Moon through those latitude lines.

As mentioned, the reduction is from about the 28 degree declination to about 18 degrees and is gradual over the nine years.

To explain the difference in practical terms, the Earth's tilt is 23½ degrees off the perpendicular from the ecliptic (the Earth's path around the Sun). The Moon's inclination of its orbit or its own tilt of five degrees to the ecliptic has a pivot point or centre for the Moon's orbit within the centre of the Earth. The high or north part of the inclination at the northern solstice position will be five degrees above the ecliptic and it will slowly circle halfway round the Earth. So when the high part of the inclination coincides with the solstice or winter position of the Southern Hemisphere on the Earth's orbit around the sun, that is when the five degree inclination is subtracted from the Earth's tilt of 23½ degrees at the northern summer solstice. The angle from the winter solstice of the Southern Hemisphere position or the ecliptic to the Earth's tilt is actually 113½ degrees, and from the inclination of the Moon at that point which is five degrees above the ecliptic, the tilt of the Earth's axis is now 108½ degrees. On Earth the latitude movement of the Moon will then be 18 degrees north and south of the Equator or a total shift of 36 degrees or five degrees inside each tropic.

After nine years the high part of the Moon's inclination again will be at the Northern Hemisphere solstice position, and on the same side of the 23½ degree tilt where it leans towards the ecliptic, or five degrees above the ecliptic and of course five degrees above the Tropic of Cancer. The angle between the tilt of the Earth's axis and the inclination of the Moon will then be 18 degrees but the movement shows on Earth a declination of 28 degrees above or below the Equator. The Moon's latitude movement, as it appears on Earth, will now be 28 degrees north or south of the Equator or a total shift of about 56 degrees.

The effect on the weather during this maximum and minimum declination of the Moon has in part been explained.

As it has been mentioned, the Moon at that time crosses the ecliptic at a point which equates to the Equator on either going up or down both for a 28 degree declination and the 18 degree declination except that the tilt or inclination is sloping in opposite directions for each of them. This point, as stated, is known as the nodes and the Moon crosses the Equator also at exactly the same point.

After approximately four and a half years, the nodal crossing of the ecliptic will occur at a point 90 degrees further on around the Moon's orbit which is also the summer or winter solstice position for either hemisphere.

At this time the Moon's declination is then reduced to about 23 degrees north or south of the Equator which is halfway between the 18 and 28 degree marks of the declination, and because it is the point where the crossings take place on the ecliptic and also halfway between each of the equinox positions, the ecliptic is also where the 23½ degree tilt of the Earth's axis is measured from.

While the Moon's orbit, when related to its inclination only, is in this position the tilt of its orbit is north and south of the equinox positions, i.e. at the March equinox position about one half of the Moon's orbit will be slanted north of that point, while the other half of the orbit will be on the opposite side and slanting south of that point which relates to the September equinox.

About nine years later the precession of the nodes will take them to the position previously occupied by each other 180 degrees away.

Then the Moon's orbit, and still referring to the inclination in respect to the ecliptic, will be south of the position of the March equinox and slanted to the north of the September equinox.

Now on going back to the subject of the oscillations, it will be noted the years given at first were 1965/66, 1974/75 and afterwards the 1955/56 years have been brought in also later in this chapter. The next nine years bring us to 1983, the year on which so much has been written on the effects of the 'Southern Oscillation'.

These are the years when the Moon's nodes were at, or near, the solstice position and which is also on the same plane as the ecliptic. These were the times when what had been termed previously as 'crazy mixed up weather' had taken place. On going back further into history and beyond the years listed here, more examples on this type of weather may be unearthed.

Every meteorologist who had a name to protect or make, wrote articles on the subject. Every unusual meteorological occurrence was connected to the 'Southern Oscillation', and no one could prove whether it was connected to the oscillation or not. It is what can be termed 'statement making', saying something for the sake of being heard, to gain an advantage over fellow meteorologists, or to see one's name up in the bright lights of prestige.

What has been written by me on the 'Southern Oscillation' probably will be ridiculed, criticised and torn to pieces because someone had not thought of the basic reasons first for that which had been responsible for the 'Southern Oscillation', and if someone cannot be first then the best alternative is to destroy another opinion.

Then back to 1955/56

I do not hold newspaper cuttings before 1962, although the newspaper files were perused and notes taken for the years 1955/56 and the following comments on those years are included.

The rough notes on weather phenomenon taken from the newspaper files of the *Waikato Times* in Hamilton, New Zealand, for the years 1955/56 are entered in sequence as they were reported. Obviously there is no guarantee that what is written here is anywhere near the amount of unusual or abnormal weather that may have occurred around the world then, but that which was reported must have been sufficiently severe for the events to be published in this country.

At the time when I gathered the material, during the 60s, El Nino as a weather phenomenon was still to be recognised and named as such in the more distant future. Again, as previously mentioned, no climatologist or meteorologist had brought together all the severe weather reports as an article and queried why the weather was misbehaving as it had been doing over that particular period.

From February 1955, reported on the 21st, blizzards west Pacific — Formosa to Kuriles. On the 22nd, blizzard in north Scotland. Drought conditions in New Zealand.

April, icy winter in England.

May, Australia gripped in severe cold spell — Tasmania to Queensland on east coast. May 27th, tornado sweeps Kansas, Oklahoma and Texas, 98 killed, 70 injured.

In July from within New Zealand — Pirongia, a mountain close to Hamilton receives snow, which is unusual at these latitudes and roads are blocked in central North Island.

August 9 report listed widespread storms in south and east Australia and on the 10th, gale force winds lash Victoria and New South Wales.

From the USA on the 21st, flooding Connecticut, Pennsylvania, 165 deaths. Floods in Peshawar, Pakistan.

September 22nd, floods Mexico, 200 dead.

October in New Zealand, Northland floods receding. On the 18th, floods in Hunterville, central North Island.

September 11th, floods threaten Pakistan and India.

September 14th, floods sweep stricken town of Volos, Greece.

In early December from New Zealand, floods receding in Northland. Violent electrical storms with hail lashing Australia reported on the 31st — this in mid-summer. December 28th report, from California — worst flooding in history, 31 inches last week. December 29th report, fresh gales lash coast of Britain.

1956: February 1st report — floods isolate Suva from the rest of Fiji. And from England worst freeze for years. February 2nd report — England frozen for third day, ships frozen in Hague. February 3rd report — big freeze easing in England, cold moving towards the Mediterranean. February 5th report — Istanbul train stranded in snow and in Tacumcari, New Mexico bus stranded in snow. February 6th report — widespread flooding southern Queensland. There were several comments on the severity of flooding in New South Wales, and Queensland, 12 inches rain over five days. Victoria to the south, sweltering in temperatures of 90°F.

December 12th report — Europe still in grip of cold. Snow in Rome, Belgrade ice up to 10 feet thick. Berlin rivers frozen. December 14th report — freeze up from Mediterranean to English Channel. December 9th report — blizzards sweeping Britain, France, Italy and Denmark. December 21st report — fresh blizzards sweeping in from North Sea. Vienna snowing; 22nd, sunshine Italy. December 23rd, 115 hours continuous frost England, longest since 1947 when 230 hours were recorded. 1947 is nine years away from 1956. December 28th report — Darwin, north-west Australia, rising floods. December 27th report — violent storms moved into east Canada. Turbulent weather from Texas to New England. Winds at 95mph moved north-east. Illinois caught full fury at night. Cold air moved on to USA, temperatures down to 22 degrees below at Minnesota. Floods southern Italy. December 29th report — from Darwin, monsoon rains swell rivers, the worst since 1914.

March 6th report — cyclone bears down on Fiji. 7th report — raging cyclone lashes northern Queensland coasts, second worst in history. 8th report — floods surging through western New York State and Port Pennsylvania, three days continuous rain and also some snow. 12th report — Queensland still flooding, a depression 183 miles west of Brisbane, capital of Queensland. Gales in Japan. 17th report — Sydney, Australia, Macquarrie River flooded. 19th report — blizzard sweeping USA. One inch of snow New York. Limited emergency in Boston. 20th report — New York heaviest snow for eight years. 26th report — Brisbane, Australia cyclone growing stronger. Three inches rain in 20 minutes.

April 7th report — depression from north Tasman bears down on to New Zealand, been raining for three days. Although not reported as a cyclone it possibly was one. 9th report — Northland, New Zealand flooding and eight inches rain, heaviest for 45 years (9 x 5).

There seems to be a gap here without any overseas news being reported, possibly an editorial decision.

May 14th — serious flooding in Victoria and New South Wales, 12 inches rain over a weekend in one area. 18th report — from Melbourne, flooding at Charlton, Australia. 19th report — cyclone east of Auckland, New Zealand. 25th report — floods in north Tasmania, worst since 1929 which is 27 years (9 x 3). 28th report — floods Northland, New Zealand, five inches in three and a half hours.

June 11th report — cyclone Sydney, winds 75-83mph. 12th report — Hong Kong and vast areas of China flooded east and central districts. The storm began on 2nd. 25th report — Sydney's biggest floods for 70 years. Melbourne coldest for six years, 45.7°F (7°C).

July 12th report — from Melbourne, floods in Murray River, biggest for 86 years. 16th report — floods Sydney, more rain expected. 22nd report — Baltimore, USA, sudden floods. 29th report — southern England, fierce storms in channel, gales 80mph. 30th report — Johannesburg, drought and cold weather causing havoc to crops, worst in living memory.

August 2nd report — typhoon hit Japan. 6th report — floods Wales, England, sunny north England.

There were no further comments on overseas weather from this point onwards, and from the earlier comment on the shortage of overseas news and to this point, only occasional references were made to overseas weather, much of it from Australia, but the earlier report on world-wide extreme weather conditions were so varied that there would seem to be very little doubt that what was reported was similar in many respects to that which had occurred in the apparent nine-ten year cycles of similar events.

It was not possible to go back another nine years to see if there were any further unusual meteorological occurrences to support the four periods already covered.

The next El Nino

If this theory is correct, the next 'oscillation', be it southern or northern, will be about 1991. 'About' is used because other factors not known could vary in which part an 'oscillation' may occur in the nine to 10 year period or even less. Perigee will occur in 1991 at or near the northern declination position from about May 1990 until January 1992, but the Moon's nodal crossings will not arrive at the solstice point until about August 1992. This situation could alter the weather pattern from that which applied previously.

Another factor is that the nine year reverse tilt of the Moon's orbit (its inclination) may have an effect in moving the atmosphere in a different manner.

In one instance, looking from one side of the equinox — say the vernal equinox position — the inclination of the Moon reduces from five degrees north of the ecliptic to five degrees south at the opposite equinox; whereas when the tilt is in the opposite direction after about nine years, or 180 degrees away, the Moon changes from five degrees south of the ecliptic at the vernal equinox to five degrees north of the ecliptic at the opposite equinox, and this change could alter the hemisphere over which an oscillation could occur. Also, perigee can take place on a different part of the Moon's orbit — a factor which will determine how the atmosphere behaves between one nine year period and the next. That is why some doubt has been offered on the hemisphere over which an 'oscillation', most likely, may take place.

Because the perigee cycle takes 8.85 years to complete and the full precession cycle of the nodes is 18.6 years, the two cycles must get out of step with one another. Twice the 8.5 years of the perigee cycle equals 17.70 or 0.9 years less than that of the precession cycle. The two cycles move in opposite directions around the Moon's orbit so a gap will be left between a nodal crossing of the precession cycle and after two complete perigee cycles, if each cycle commences from a nodal crossing as a common starting point.

For each successive precession cycle completed, the double perigee cycle will be further away from the original starting point of the nodal crossing.

The perigee cycle has been doubled so that it can be compared more sensibly with a single precession cycle but as the 'oscillation' is nearer nine years, then the single perigee cycle is the one where the most interest lies.

In June of 1955 when the crossings were near the solstice positions, perigee occurred at the northern declination point and perigee remained near or at the northern declination until about August 1956.

In January 1965 the crossings were past the solstice position, while perigee was at the northern declination position and remained near there until February 1966 before commencing to move towards the Equator where perigee took place about November 1966.

In September 1973 the nodes were crossing at or near the solstice position and perigee took place at the northern declination and remained near there until about July 1974 before moving towards the Equator and occurring about there in March 1975.

In January 1982 the crossings were approaching the solstice position and passed that position by December 1983. The Moon was at perigee in the northern declination position throughout this period.

In each instance when the oscillations occurred, the Moon's nodes were near the solstice positions, but the Moon was also at perigee at or near the northern declination point, which in itself appears to be a significant factor in the true cause of the oscillations. But it must be remembered that the inclinations of the Moon's orbit was leaning in opposite directions in each period which could alter the weather patterns for each of the years involved in the oscillations.

The opposite situation could occur when the Moon will be at perigee at the southern declination position and the nodes near the solstice point of the Moon's orbit. This does occur but as my records do not contain sufficient weather records to tie in with such a situation, its effect cannot be commented upon. No doubt it will be researched on some future occasion.

It may well be that the unusual weather experienced in the quoted examples could be the only unusual weather situations likely to occur in the present era because the Moon at perigee near a north or south declination point, and the nodes near the solstice, could have a fairly lengthy period between events.

On reading articles on the 'Southern Oscillation', comments do remark on the possible number there may have been this century. One stated about six and another about 12. But if this theory of mine is correct, then the number would depend on how many times perigee occurred at the north or southern declination points while the nodes were crossing close to the solstice position.

It has been said that the years between 'oscillations' can be two or three to five years, but factually, some of the events meteorologically may not be true 'oscillations', just said to be one because the true reason for their development has not been fully understood.

A letter to the editor of a NZ newspaper, while this book was being written, making the point that meteorologists were blaming every unusual weather phenomena, which could be a normal weather change, on an 'oscillation' brought forth a number of articles supporting their contention that the 'unusually mild winter of 1987' was the result of an 'oscillation'. With might on their side how could one lone voice satisfactorily dispute this viewpoint?

Another point to dwell upon could be that, at different times, it may be that the 'oscillations' in the 9-10 year cycle (approximately) may not be as pronounced as they seemed to have been in 1983; in other words the cycle could have an increasing tempo within the cycle and return to a more benign influence after reaching a peak.

The nodal crossings near this time which give this 'oscillation' effect do not seem to occur exactly at the solstice positions but sometimes a little past the solstice, which makes the period nearer 10 years than nine although the years affected may overlap both years and the position of perigee seems to be a contributing factor.

On the 20th October 1983, Dr Harold Ward a meteorologist attending a conference in Wellington, New Zealand, said the 'Southern Oscillation' could re-occur from the alignment of the major planets in a 60 degree segment of the sky, which happened only once in 179 years or thereabouts.

The comment is reminiscent of that by John Gribbin and Stephen Plagemann who jointly wrote *The Jupiter Effect*, a book prophesying severe earthquakes in California using exactly the same argument. When the alloted time arrived without the severe earthquakes, or even any, they said they really did not believe it themselves.

But they wrote the book which was widely publicised and caused considerable apprehension within California at that time.

The El Nino or southern oscillation of 1982-83 has for some years been the excuse for predicting any unusual weather change after that event. Now a new title has been given to what is referred to as the "positive phase of the Southern Oscillation" over the summer of 1988-89 and called La Nina. It is said to be the opposite of El Nino and is stated to be the reason for the drought conditions along the east coasts of the South Island of New Zealand during that summer when most of the weather systems crossed the country from the west and with rain on that coast, but dry to the east.

Is it a coincidence that the Moon is just declining from the 28 degree maximum declination? Strangely 1969-70 was also dry as were parts of New Zealand in 1950, 1932 and also 1914 when in each of those years the Moon was at the 28 degree declination stage.

After nine years the Moon will be at the 18 degree declination position. What name will be given to the weather pattern occurring at that time? The weather will be more unsettled then so a name appropriate to those conditions will be required.

Giving the various weather conditions a name only highlights the fact that meteorologists do not understand why the weather behaves in the manner it does during the differing declination sections which broadly are broken into four parts: when the Moon is at the 28 degree maximum declination, now called La Nina; descending to 23½ degrees, (El Nino); then to 18 degrees minimum declination (as yet unnamed); and ascending to 23½ degrees (also El Nino); and up to 28 degrees again. But remember, of the two El Nino positions, each one has the Moon's inclination to the ecliptic sloping in opposite directions so the effect from each El Nino could be different.

10: Hurricanes, cyclones, typhoons

Cyclones of the southern Pacific Ocean, hurricanes of the West Indies, or typhoons of Japan; whatever the name used, they bring fear to the population affected by them. Even cyclones of a modest size can cause vast destruction and loss of life, but cyclones may vary in size from 200 to 1000 miles in diameter.

Because the author belongs to an area where the term 'cyclone' is commonly used, that term will be maintained in this chapter.

How they begin to develop is still in the realms of conjecture. Much is known about them, that a cyclone needs a sea temperature of at least 26/27°C before forces begin to act on a thick blanket of moist air. Some authorities say that the spin of the Earth on its axis sets the air mass circulating, others give the credit to the Coriolis Force.

Cyclones commence life somewhere between five degrees and 20 degrees of the Equator and they always move away from the Equator. One forming north of the Equator will never cross over to the Southern Hemisphere, nor vice versa.

The higher the water temperatures, the more violent the cyclones may become. And while the cyclone remains over water with the temperature required to maintain its fury, it is dangerous to anything in its path.

A cyclone will diminish in intensity only when the water temperature drops below generation level, or when the cyclone moves onto land where it will lose its supernatural strength through loss of heat.

A cyclone has an 'eye' in the centre of the whirling mass of moist air where barometric pressures are very low compared to pressures further away toward the edge. In the eye it is relatively calm, but at the sides of the eye the wall of moist air circulates with increasing wind speed towards the outer edge of the cyclone and can reach up to or well over 160kmph (100mph).

The eye sucks in masses of moist air from the surrounding area and this in turn assists in maintaining the energy within this massive and destructive storm system.

While over the sea, the low pressures within the eye raise the level of the sea by many feet which results in flooding when the cyclone moves on to a coast or over tropical islands. This is over and above the torrential rain which accompanies the cyclone. Coupled to the screaming destructive winds, cyclones under any guise are a force to be feared by those unfortunate enough to be affected by them.

Cyclones rotate in opposite directions over each hemisphere. Over the Northern Hemisphere their rotation is anticlockwise, while clockwise over the south. Over each hemisphere the cyclones will drift westward at first, but slowly arc away from the Equator and eventually to the north-east over the Northern Hemisphere and to the south-east over the Southern Hemisphere. Should a cyclone remain active as it moves toward cooler waters, the drift will slowly change direction to the east, but by this stage it will have moved a sufficient distance from the Equator to be into cooler waters where it will have become a very active depression but slowly losing its previous intensity.

The eastward drift of a cyclone does not occur very often because the seas, where most cyclones develop, are not large enough to allow this change of direction to occur. The cyclones usually approach a land mass and lose their intensity before they can make this drift to the east.

An exception to this general rule is over the larger area of the Pacific Ocean south of the Equator, where cyclones, after completing their westward movement from the mid-Pacific, circling towards the south-east can begin to drift east if they have not reached a land mass such as Australia during their westward drift.

The many tropical islands in the west Pacific (such as Fiji, Tonga, Samoa and Cook Islands) are visited by cyclones every now and again with disastrous results. Many of the tropical islands are low lying and the sea surges can cause huge damage. The winds destroy the lightly built houses and uproot trees, most of them coconut palms on which the population relies for foreign exchange and income for the individual.

This is but a generalisation on cyclonic behaviour. Anyone interested in a more detailed explanation on the formation and the practical aspects of cyclones will find many books available which will assist them in gaining further knowledge on the birth, life and decay of cyclones.

The object of this book is to offer a lunar connection between the atmospheric movement and weather generally, in whatever form the meteorological behaviour may manifest itself, whether it be tornadoes, snow storms, anticyclones, depressions, electrical storms or cyclones.

Cyclones — synonymous for hurricanes and typhoons and other localised names in various parts of the world — are also affected by the Moon's gravitational attraction. In the past, as now, it has always been acknowledged that there are 'seasons' for the development of cyclones, June to September in the north and December to March over the Southern Hemisphere, these being the normally warm summer months over each hemisphere.

However, some cyclones do occur outside these months every now and again.

We know that heat is a necessary ingredient before cyclones will begin to develop and the present thinking is that the spin is imparted either by the Coriolis Force or through the spin of the Earth on its axis, depending on which school of thought to which one belongs.

But as has been mentioned previously, if the Earth was without a moon, it would be unlikely that a cloud mass, which would always be facing the Sun, would allow sufficient heat to develop for a cyclone to form. Therefore neither the Coriolis Force nor the Earth's spin could inhibit the formation of a cyclone. But the Earth does have a moon (and by its action the cloud mass is broken into smaller cloud units around the Earth), and because of the Moon it does not necessarily follow that either of the other two forces do not assist in causing a cyclone to develop. There is no positive proof for or against such a proposal, just an opinion voiced by meteorologists gained only from the knowledge available to them at the present time.

Mention has to be made again that the Moon must offer some gravitation attraction to our atmosphere and because Earth is spinning on its axis inside the Moon's orbit, the Moon also could impart an action through this gravitational attraction, on the atmosphere which may start a potential cyclone circulating.

It may be argued that this suggestion is not a proven fact and this point is agreed upon. But the present view is also only an opinion voiced by 'experts' and without any positive proof that their opinion has a more valid base than this one now being raised by myself.

When one looks at the circulation of cyclones — anticlockwise in the Northern Hemisphere and clockwise in the Southern Hemisphere — one fact becomes obvious; cyclones in each hemisphere, and looking at the circumferences nearest the Equator, do circulate towards the east and into the direction of the Earth's spin on its axis. Put another way, the direction of the Earth's rotation and the rotation of the cyclone's edge, which is closest to the Equator, are both moving in the same direction.

Now, if the Earth's spin was the only cause which allowed a potential cyclone to rotate, then the present direction of rotation could only be so if there was some form of drag somewhere on the atmosphere, such as gravitation attraction. For without drag the atmosphere would keep pace with the Earth's spin, there being nothing to prevent that happening.

If the Coriolis effect was the only cause transferring rotation to a cyclone, then because the edge of the cyclone nearer the tropics was on a latitude with a smaller circumference around the Earth, it seems likely that the direction of rotation would be to the east on that edge nearer the tropics because that edge would try to move faster, and therefore opposite to the present rotation. And, as a cyclone drifted further away from the Equator, the circumference of the Earth would diminish even further accentuating that probable rotation.

This is an area for speculation, but as cyclones have a diameter of up to 1600km (1000 miles), the edge furthest from the Equator must be affected by the diminishing circumference of the Earth at that point.

Cyclones have a 'season' and this coincides with the Sun being over a particular hemisphere, and as the summer of that hemisphere is extended into the autumn, the sea temperatures slowly build up to a maximum rising to that critical point where cyclones are likely to develop.

While the Sun is over one hemisphere it is unusual for cyclones to develop over the other, excepting within a week or two of the equinox, and when cyclones do develop under these circumstances, they usually form over the hemisphere the Sun has just vacated.

This means that, in most instances, the Sun is nearer the tropics when the cyclones begin to form closer to the Equator. The gravitational attraction and drag by the Sun on the atmosphere could assist in imparting a westerly drift to that edge away from the Equator and also assist in giving the circulatory pattern to the cyclones.

As the autumn season approaches, the last quarter moon will also move into that hemisphere and be at a latitude above a developing cyclone, again assisting in imparting a westerly drift to the edge nearer the tropics.

The anti-clockwise spin of cyclones over the Northern Hemisphere and the clockwise spin over the Southern Hemisphere may also contribute to the cyclones taking a line which allows them only gradually to drift away from the Equator during their westward track. In other words, the cyclonic mass spinning as it does, would create friction on the sea below and this would have a tendency to steer the cyclones towards the Equator.

This action would explain why the cyclones maintain a mainly westward movement at first and, as the cyclones do move further away from the Equator into higher latitudes, the swing away from the Equator becomes more pronounced because of the Coriolis effect, which in this instance could be the correct force operating on the cyclones.

It is also interesting to note that the total mass of the cyclones in each hemisphere almost copy the circulatory direction which is ultimately to the east, the same direction as the outer edge of anticyclones, i.e. the edge away from the Equator, a point not mentioned in meteorological writings in the past.

Most of the cyclones develop in the period between the summer solstice and the equinoxes, which is when the first or last quarters of the Moon are leaving the normal equinox position, i.e. the point where the Sun would be situated when at the equinox position. During the summer of the Northern Hemisphere, the Sun would be at the solstice position about 22 June each year. The phases of the Moon are variable at or near this date, and may be near the last quarter, new, or first quarter. But in general, when the Sun is at or near its solstice position, the new moon is positioned in the same direction as the Sun. As the Moon moves away from the summer solstice position, the first quarter is usually near the autumn equinox. The full moon is opposite the Sun and the last quarter near the vernal equinox.

For the Southern Hemisphere summer solstice, the quarter moons are in the opposite positions, and while new and full moons are in the same position in relationship to the Sun, they also are in opposite positions to those they occupied over the Northern Hemisphere.

Although the details of the relative positions of the phases of the Moon while the Sun is at the solstice position may not be fully understood by many, they are given so that any statement on the possible effect by the moon on cyclonic behaviour can be followed by anyone who may wish to do so.

The main point, in offering this detail, is to stress that as the Earth is advancing on its orbit from the summer solstice position in either hemisphere and towards the equinox, a last quarter moon will move up into the same hemisphere as the Sun, which means obviously that the Sun and Moon are both over the same hemisphere.

It has been mentioned previously that a full moon will give a lower daytime atmospheric tide, while a last quarter moon also will have a lower atmospheric tide from about midday onwards and into the evening, which means that more heat will be generated on the hemisphere experiencing summer.

We know that warm sea temperatures are required to allow cyclones to develop and it must be a prerequisite that these conditions exist when the cyclones do form.

Cyclones and the Moon's influence

Since about 1962 newspaper cuttings have been filed away regularly so that at the time of writing, about 260 cuttings on cyclones have been accumulated. These have been sorted into the hemispheres the cyclones were over, and also into moon phases at that time. The moon phases have been further sorted into the hemisphere the Moon occupied at the time of the cyclone.

Another sorting involved placing the cyclones into two separate sections of the lunar month which was; one section to be between perigee and apogee and the other section between apogee and perigee.

As has been mentioned, the Moon has a more rapid movement in its orbit around the Earth while at perigee and a lesser orbital speed while at apogee.

Now, the break between perigee and apogee and apogee back to perigee is not always an equal split. The time taken between each section on the orbit is variable and it has differing times from just over 380 hours down to just over 280 hours. (Hours have been used in preference to days to gain greater accuracy.) The number of hours taken for each section on the orbit changes to more hours or less hours each lunar month, so that the perigee to apogee section on the orbit may be say, 380 hours in one month reducing to 280 hours in about three or four months.

There is a pause when the 280 hours that the perigee/apogee section has been reduced to, and the increased 380 hours of the apogee/perigee section. Both remain more or less constant for another month before the hours increase again in the first section on the orbit and with the hours then reducing in the other section, which will explain the three or four months mentioned.

The end result of this change is that the orbital speed of the Moon in the shorter 280 hours will be declining at a faster rate between perigee and apogee and increasing at a slower rate from apogee to perigee because the hours in this section will then be near 380.

The reverse occurs when the hours are 380 between perigee and apogee and the decline in orbital speeds will be slower, but the increase in speed between apogee and perigee at 280 hours will be much quicker.

An explanation for anyone interested in this phenomenon from an astronomical point of view is; the apogee section of the Moon's orbit advances steadily around the orbit for its 8.85 year cycle by an average of about three degrees each month but the position of perigee varies according to the movement of both the new and full moons, it will advance as apogee will do but as either the new or full moons approach the perigee section of the orbit, perigee will rebound a few degrees back along the orbit path which action reduces the hours between apogee and perigee to the 280 hours. Perigee will again advance along the orbit at a faster pace than apogee until the 380 hours have been reached when the new or full moon will again overtake perigee on the orbital path. This routine will continue to be repeated for the 8.85 year cycle.

The above explanation refers to altering orbital speeds and can be related to changing rates of movement through the latitudes which are taking place by the Moon at the same time. The time taken for the Moon to move from say, the southern declination through to the northern declination and back again, will relate to those changing orbital speeds.

So it must follow that if it is acknowledged that the Moon does disturb the atmosphere, it must do so at a changing rate.

It may be that it is in this area that cyclones may be encouraged to develop, possibly during the shorter number of hours taken by one section or the other. But an increasing orbital speed in the short 280 hours of apogee to perigee could be the catalyst to speed up the atmospheric movement to enable cyclones to form; or it could be said a slower build up of orbital speed between apogee to perigee when at 380 hours offers more time for the cyclones to develop.

It must be remembered that the longer or shorter hours apply in their turn to each section in the orbit.

Then another factor will come into focus. If one or the other of the above points is valid, and it may be that the analysis to follow is in reverse, the faster or slower orbital speeds may need to be at the same time as when the Sun is moving from the solstice position and towards the Equator, so that sufficient heat is generated to set the processes in action to form cyclones.

There are many variables to be taken into account, or coupled to the changing orbit speeds of the Moon, and not the least is, at what phase is the Moon and how many combinations of full or last quarter moons will there be when conditions could be favourable — if favourable is the correct word under the circumstances — between say, the position of the summer solstice and the equinox.

The period could be extended, at times, from say an earlier part of the lunar month commencing before the Sun is at the solstice to a later part of the Moon's orbit just after the Sun was at the equinox. Cyclones may be less frequent outside the normal 'season' but nevertheless they have been recorded.

The cyclones used by me for analysis are from 1962 to early in 1986 and as mentioned, they are not necessarily anywhere near the total number occurring in that period, but in New Zealand the newspapers publish more overseas news than seems to be the case in other countries, which will make the list fairly comprehensive.

However, the sample of 260 cyclones will offer a reasonable balance in their various sections.

No attempt has been made to break the cyclones into some of the other divisions that have been suggested. It would be too time consuming for me and the conclusions may not be as accurate as one would like. The various avenues that can be followed, and possibly some which have not been mentioned, will keep researchers going for some time. But there seems little doubt that the conclusions reached by myself, also will be reached by the said researchers. In fact, the results will open up new horizons in meteorological knowledge.

An analysis of cyclones into moon phases, etc.

The analysis which follows has been broken into moon phases and the direction the Moon was moving towards its maximum declination and the hemisphere the Moon was over at the time of the various cyclones, the Moon's movement when between apogee and perigee and perigee to apogee and in both maximum and minimum hours involving each movement.

The breakdown also covers the hemisphere which the Sun was over when the cyclones were active.

Unfortunately there was no information available giving the total time the cyclones were active, so only the newspaper dates of the event have been used, although allowances were made where needed.

The time when the Moon may have been at its maximum or minimum declination or anywhere in between was not taken into account, so the Moon may have been anywhere between 18 and 28 degrees north or south of the Equator. This is one area where viable analysis could be beneficial.

Abbreviations will be used to conserve line space. A — apogee, P — perigee, SH — Southern Hemisphere, NH — Northern Hemisphere, Cyc — cyclone. Maximum or minimum = hours.

Sample 1

Phase new over NH going north.

| A to P at maximum Sun SH | Nil Cyc SH | Nil Cyc NH |
| Sun NH | 1 Cyc SH | 4 Cyc NH |

A to P at minimum	Sun SH	Nil Cyc SH	Nil Cyc NH
	Sun NH	2 Cyc SH	3 Cyc NH
P to A at maximum	Sun SH	1 Cyc SH	Nil Cyc NH
	Sun NH	1 Cyc SH	Nil Cyc NH
P to A at minimum	Sun SH	1 Cyc SH	Nil Cyc NH
	Sun NH	1 Cyc SH	Nil Cyc NH

Over 14 samples, seven in SH, seven in NH

Sample 2

Phase first quarter over NH going north

A to P at maximum	Sun SH	5 Cyc SH	1 Cyc NH
	Sun NH	1 Cyc SH	2 Cyc NH
A to P at minimum	Sun SH	5 Cyc SH	3 Cyc NH
	Sun NH	Nil Cyc SH	Nil Cyc NH
P to A at maximum	Sun SH	8 Cyc SH	3 Cyc NH
	Sun NH	Nil Cyc SH	1 Cyc NH
P to A at minimum	Sun SH	Nil Cyc SH	1 Cyc NH
	Sun NH	Nil Cyc SH	2 Cyc NH

Over 32 samples, 19 in SH, 13 in NH

Sample 3

Phase full over NH going north

A to P at maximum	Sun SH	1 Cyc SH	2 Cyc NH
	Sun NH	1 Cyc SH	6 Cyc NH
A to P at minimum	Sun SH	2 Cyc SH	2 Cyc NH
	Sun NH	Nil Cyc SH	1 Cyc NH
P to A at maximum	Sun SH	1 Cyc SH	Nil Cyc NH
	Sun NH	Nil Cyc SH	5 Cyc NH
P to A at minimum	Sun SH	Nil Cyc SH	6 Cyc NH
	Sun NH	Nil Cyc SH	2 Cyc NH

Over 29 samples, five in SH, 24 in NH

Sample 4

Phase last quarter over NH going north

A to P at maximum	Sun SH	1 Cyc SH	2 Cyc NH
	Sun NH	Nil Cyc SH	6 Cyc NH
A to P at minimum	Sun SH	1 Cyc SH	Nil Cyc NH
	Sun NH	1 Cyc SH	8 Cyc NH
P to A at maximum	Sun SH	2 Cyc SH	Nil Cyc NH
	Sun NH	Nil Cyc SH	3 Cyc NH
P to A at minimum	Sun SH	3 Cyc SH	Nil Cyc NH

	Sun NH	1 Cyc SH	3 Cyc NH

Over 31 samples, nine in SH, 22 in NH

Sample 5

Phase new over NH going north
No cyclones in SH or NH

Sample 6

Phase first quarter over NH going south

A to P at maximum Sun SH	3 Cyc SH	1 Cyc NH
Sun NH	Nil Cyc SH	Nil Cyc NH
A to P at minimum Sun SH	Nil Cyc SH	Nil Cyc NH
Sun NH	2 Cyc SH	Nil Cyc NH
P to A at maximum Sun SH	1 Cyc SH	Nil Cyc NH
Sun NH	2 Cyc SH	Nil Cyc NH
P to A at minimum Sun SH	2 Cyc SH	Nil Cyc NH
Sun NH	Nil Cyc SH	2 Cyc NH

Over 13 samples, 10 in SH, three in NH

Sample 7

Phase full over NH going south

A to P at maximum Sun SH	Nil Cyc SH	Nil Cyc NH
Sun NH	Nil Cyc SH	Nil Cyc NH
A to P at minimum Sun SH	3 Cyc SH	1 Cyc NH
Sun NH	Nil Cyc SH	Nil Cyc NH
P to A at maximum Sun SH	1 Cyc SH	1 Cyc NH
Sun NH	Nil Cyc SH	Nil Cyc NH
P to A at minimum Sun SH	1 Cyc SH	Nil Cyc NH
Sun NH	Nil Cyc SH	Nil Cyc NH

Over seven samples, five in SH, two in NH

Sample 8

Phase last quarter over NH going south

A to P at maximum Sun SH	Nil Cyc SH	Nil Cyc NH
Sun NH	Nil Cyc SH	1 Cyc NH
A to P at minimum Sun SH	Nil Cyc SH	Nil Cyc NH
Sun NH	Nil Cyc SH	Nil Cyc NH
P to A at maximum Sun SH	Nil Cyc SH	1 Cyc NH
Sun NH	Nil Cyc SH	2 Cyc NH

| P to A at minimum Sun SH | 1 Cyc SH | 2 Cyc NH |
| Sun NH | Nil Cyc SH | Nil Cyc NH |

Over seven samples, one in SH, six in NH

Sample 9

Phase new over SH going south

A to P at maximum Sun SH	Nil Cyc SH	1 Cyc NH
Sun NH	Nil Cyc SH	3 Cyc NH
A to P at minimum Sun SH	1 Cyc SH	1 Cyc NH
Sun NH	Nil Cyc SH	Nil Cyc NH
P to A at maximum Sun SH	Nil Cyc SH	1 Cyc NH
Sun NH	Nil Cyc SH	1 Cyc NH
P to A at minimum Sun SH	2 Cyc SH	1 Cyc NH
Sun NH	Nil Cyc SH	1 Cyc NH

Over 12 samples, three in SH, nine in NH

Sample 10

Phase first quarter over SH going south

A to P at maximum Sun SH	Nil Cyc SH	1 Cyc NH
Sun NH	Nil Cyc SH	2 Cyc NH
A to P at minimum Sun SH	1 Cyc SH	4 Cyc NH
Sun NH	Nil Cyc SH	7 Cyc NH
P to A at maximum Sun SH	Nil Cyc SH	4 Cyc NH
Sun NH	Nil Cyc SH	3 Cyc NH
P to A at minimum Sun SH	Nil Cyc SH	1 Cyc NH
Sun NH	Nil Cyc SH	15 Cyc NH

Over 38 samples, one in SH, 37 in NH

Sample 11

Phase full over SH going south

A to P at maximum Sun SH	3 Cyc SH	Nil Cyc NH
Sun NH	Nil Cyc SH	2 Cyc NH
A to P at minimum Sun SH	3 Cyc SH	Nil Cyc NH
Sun NH	Nil Cyc SH	3 Cyc NH
P to A at maximum Sun SH	2 Cyc SH	Nil Cyc NH
Sun NH	1 Cyc SH	3 Cyc NH
P to A at minimum Sun SH	5 Cyc SH	Nil Cyc NH
Sun NH	2 Cyc SH	Nil Cyc NH

Over 24 samples, 16 in SH, eight in NH

Sample 12
Phase last quarter over SH going south

A to P at maximum	Sun SH	7 Cyc SH	3 Cyc NH
	Sun NH	Nil Cyc SH	Nil Cyc NH
A to P at minimum	Sun SH	3 Cyc SH	2 Cyc NH
	Sun NH	Nil Cyc SH	3 Cyc NH
P to A at maximum	Sun SH	5 Cyc SH	2 Cyc NH
	Sun NH	Nil Cyc SH	Nil Cyc NH
P to A at minimum	Sun SH	6 Cyc SH	1 Cyc NH
	Sun NH	Nil Cyc SH	Nil Cyc NH

Over 32 samples, 21 in SH, 11 in NH

Sample 13
Phase new over SH going north

A to P at maximum	Sun SH	Nil Cyc SH	Nil Cyc NH
	Sun NH	1 Cyc SH	Nil Cyc NH
A to P at minimum	Sun SH	Nil Cyc SH	Nil Cyc NH
	Sun NH	Nil Cyc SH	Nil Cyc NH
P to A at maximum	Sun SH	Nil Cyc SH	Nil Cyc NH
	Sun NH	Nil Cyc SH	1 Cyc NH
P to A at minimum	Sun SH	Nil Cyc SH	Nil Cyc NH
	Sun NH	Nil Cyc SH	Nil Cyc NH

Over two samples, one in SH, one in NH

Sample 14
Phase first quarter over SH going north

A to P at maximum	Sun SH	Nil Cyc SH	Nil Cyc NH
	Sun NH	Nil Cyc SH	1 Cyc NH
A to P at minimum	Sun SH	Nil Cyc SH	1 Cyc NH
	Sun NH	Nil Cyc SH	2 Cyc NH
P to A at maximum	Sun SH	1 Cyc SH	2 Cyc NH
	Sun NH	Nil Cyc SH	4 Cyc NH
P to A at minimum	Sun SH	1 Cyc SH	Nil Cyc NH
	Sun NH	Nil Cyc SH	3 Cyc NH

Over 15 samples, two in SH, 13 in NH

Sample 15
Phase full over SH going north
No cyclones over SH or NH

Sample 16
Phase last quarter over SH going north

A to P at maximum	Sun SH	Nil Cyc SH	Nil Cyc NH
	Sun NH	Nil Cyc SH	1 Cyc NH
A to P at minimum	Sun SH	Nil Cyc SH	Nil Cyc NH
	Sun NH	Nil Cyc SH	Nil Cyc NH
P to A at maximum	Sun SH	1 Cyc SH	Nil Cyc NH
	Sun NH	Nil Cyc SH	Nil Cyc NH
P to A at minimum	Sun SH	1 Cyc SH	Nil Cyc NH
	Sun NH	Nil Cyc SH	1 Cyc NH

Over four samples, two in SH, two in NH

Conclusions

On going through the previous analyses there are several obvious parallels with the same moon phases but in opposite hemispheres, in the examples offered. The Moon will be moving either towards the Equator or away, so that the situation is the same.

With the Moon at new over the Northern Hemisphere and going north, in sample 1, there were a total of 14 cyclones.

In sample 9 with the new moon over the Southern Hemisphere and going south, the cyclones total 12.

The new moon of sample 5; there are no cyclones and in its opposite position, sample 13, there were two cyclones.

The Moon in the first quarter in sample 2; there were more cyclones in the Southern Hemisphere, 19 against 13 in NH. Total 32.

In sample 10 the situation was reversed with one cyclone in the Southern Hemisphere and a massive 37 in the Northern Hemisphere, total 38.

In the first quarter of sample 6, there were 10 cyclones in the Southern Hemisphere and three in the Northern Hemisphere, total 13.

In the opposite sample 14, there were two cyclones in the Southern Hemisphere and 13 in the Northern Hemisphere, total 15, which figures are almost reversed.

In the full moon of sample 15, there were no cyclones in either hemisphere.

In the last quarter moon sample 4, there were nine cyclones in the Southern Hemisphere and 22 in the Northern Hemisphere, total 31.

In the opposite sample 12, there were 21 cyclones in Southern Hemisphere and 11 in the Northern Hemisphere, which numbers are almost reversed.

In the last quarter phase of sample 8, there was one cyclone in the Southern Hemisphere and six in the Northern Hemisphere, total seven.

In the opposite sample 16, there were two cyclones in the Southern Hemisphere and two in the Northern Hemisphere, total four.

The new moon of sample 5 and the full moon of sample 15, when no cyclones were reported, have one point in common: the new moon was over the Northern Hemisphere and going south towards the Equator.

The full moon was over the Southern Hemisphere and going north towards the Equator.

What the astronomical significance would be is not known, but there must be a common denominator which will be left for someone more knowledgeable than myself to solve.

Although there were no cyclones mentioned in the newspaper cuttings used, it is acknowledged that there may have been an odd cyclone somewhere, but it may not have been large or troublesome. The fact at issue is that the number would be small if any cyclones were about, and related to those two samples.

One unexpected result occurred in the samples 2 and 10 where the moon phase was in the first quarter going north and away from the Equator over the Northern Hemisphere, and over the Southern Hemisphere going south away from the Equator also: the cyclones total 70 which is a high percentage of the 260 samples given. Of the 70 samples, 50 occurred over the Northern Hemisphere. This suggests that with the Moon in that position it would be a most dangerous time for yachts or any other small craft to be at sea, or if one must be at sea, it would be advisable to stay close to a port or safe harbour. To undertake long voyages away from land would seem to be unwise.

From the newspaper cuttings used to compile this analysis, there were 158 cyclones over the Northern Hemisphere and 102 over the Southern Hemisphere, and although the numbers may have been more, it does seem likely more cyclones develop over the Northern Hemisphere than the Southern Hemisphere. This may come about because of the larger land masses of the north, which could lift the air temperatures to higher levels than over the south, which is largely covered by oceans.

Land heats more quickly than the sea and this extra heat may be transferred to the seas of the Northern Hemisphere more quickly, resulting in more cyclones.

The large areas of the South Pacific and the southern oceans, including the South Atlantic, will heat up only slowly, and this fact may also explain why cyclones seldom develop over the mid-South Pacific regions and apparently never over the South Atlantic.

An analysis done on the number of cyclones in each section of the perigee/apogee-apogee/perigee where the number of hours in each section change, showed no particular advantage from one to the other, which means that the shorter or longer hours gave a similar number of cyclones in each section.

In A to P at maximum hours (380) 63 cyclones
In A to P at minimum hours (280) 65 cyclones
In P to A at maximum hours   64 cyclones
In P to A at minimum hours   68 cyclones

Unfortunately it was impossible to identify completely if any cyclones overlapped from one section into another, or if cyclones developed and died within a particular section, so the analysis is mainly arbitrary.

When the maximum and minimum hours for each sections were added together and related to the hemisphere which the Moon occupied at the time of the cyclones, the figures were similar.

A to P at maximum and minimum   Moon SH   64 cyclones
P to A at maximum and minimum   Moon SH   72 cyclones
A to P at maximum and minimum   Moon NH   64 cyclones
P to A at maximum and minimum   Moon NH   60 cyclones

As it will be noted, these figures are inconclusive in that they do not prove that one section will spawn more cyclones than another section.

However, whatever other factors may influence the development of cyclones, the analyses on the Moon's effect cannot be ignored; it is too conclusive.

One other factor which has not been considered in the analysis has been, over which hemisphere perigee has been situated. It may well be that the Moon at perigee at whatever phase and situated over either the northern or southern-most point in its declination could have an influence on the development and intensity of a cyclone. This is especially if perigee occurs at the same time as a night moon which could be in the late phase of the first quarter, the full moon or the early stages approaching the last quarter — all of which will give a daytime low atmospheric tide and of course a greater heat input during the day.

It was felt that an analysis such as this would be better undertaken over a longer time span and it could incorporate the maximum and minimum declination periods as well.

The years used in the analysis done would take in about three perigee cycles but only about one and a half precession cycles which seems insufficient in time to offer a satisfactory answer to these comments.

Meteorologists who have been taught the 'orthodox' answers to cyclonic development will have to do a re-think on this subject.

Meteorological text books will need to be rewritten for what use is a text book if the basics are not correct?

11: Climate

Climate is history of the weather, and the weather we have now, will be history as it recedes into the distant past to become in turn, climate.

Everybody sees climate differently and most remember the best weather in preference to the worst, but there comes a time when there is no memory left, only written records.

Before that there is what can be interpreted from the surroundings, from Antarctica, glaciers, lakes or sea beds. Tree rings are also a source for obtaining some details of past weather, especially the very old trees which have been growing for thousands of years.

One such tree, the Bristlecone pine, growing high in the White Mountains of California, has rings which can be counted back to 2000 BC or more.

Although tree rings may assist in climatic investigation, the trees are very local and what they tell us would be difficult to use for world-wide understanding of the past climate, especially when one group of trees in one area is all that is available for going back say, 4000 years.

Cores from the sea bed are another source for attempting to trace the history of the world's climate, but again how useful is such a small area from which the cores were taken for world-wide understanding of climate? Cores from the sea bed may take one back a few million years, but where was that part of the sea bed all those millions of years ago?

It may be part of, say, the Pacific today, but where was the Pacific then? Where was the Equator then or the poles? Where were the continents?

Sometimes maps are drawn of the Earth as someone imagined it to have looked in the geological past.

At one stage it is claimed that Australia, New Zealand, Antarctica, South America and Africa were all linked in a land mass referred to as Gondwanaland. The land mass is shown as being over the Southern Hemisphere and centred near the South Pole where it was supposed to be at that time. The Atlantic Ocean is non-existent with Africa and the South American continents joined together. Another land mass called Laurussia over the Northern Hemisphere which included Europe and Asia joined to North America with Greenland squeezed in between them.

All the land masses are shown on one side of the Earth. What was on the other side? Oceans? Somehow no one ever refers to the other side of the Earth.

Is it possible that the Earth was smaller in that far off geological past. If the Earth had been smaller, it would explain many of the puzzling thoughts on the type of animal life which was in existence then.

There could be an explanation or answer to the reasons for the continental drift, as a theory, within this statement also.

The gravitational forces on Earth could have been much stronger than now, and the atmospheric pressures also greater than at present. Further reference will be made to this aspect elsewhere.

The comment on the existence of the Gondwanaland and Laurussia continents and the land masses which comprised them has been made to show the futility of taking seabed cores which show how the climate may have been some 45 million years ago, when it cannot be said exactly where that portion of the seabed was situated.

The climate would change gradually over any one of those land masses, as they drifted apart, which could be north, south, east or west of where they had been situated originally.

The north and south poles do, and have, moved over lengthy periods and even if the shift was of only a few degrees, the position of the Arctic and Antarctic would alter, and of course all the tropical lands in between the tropics would lie at a different angle to that of the present time. This in turn would gradually alter the climate in many areas.

The shift of the poles could be in any direction and if, for example, the South Pole was to shift towards Australia and New Zealand, the climate of the southern area of these countries would become much colder.

The North Pole would then be closer to the British Isles and Europe. The Arctic Circle could be placed nearer Scotland. The weather would be colder there, but milder in Alaska.

Any changes such as these, would occur over tens of thousands of years, or even millions of years.

Precession of the poles

Other climatic changes relate to the precession of the poles by describing a small circle as seen from space, which means that the summers and other seasons slowly change their position on the calendar so that eventually the summers change position with the winters. As far as we are concerned, we will never notice this alteration in the seasons for it will take about 13,000 years, half the cycle, for the summer season to take over the winter position on the calendar. The shift amounts to about 20 minutes a year or one hour in three years or one day in about 72 years. The vernal equinox in 1901 occurred on 22 March and on the 21st in 1973. It will take approximately 2230 years for the equinox to shift to near the 19 February. This will be somewhere near the year 4216.

Alteration in the tilt of the Earth's axis

Another alteration will take place in the amount of tilt in the Earth's axis which at present is 23½ degrees off the perpendicular. The variation is from a minimum about 21.8 degrees to a maximum of 24.4 degrees over a period of 40,000 years.

At the present time the Earth's tilt is decreasing so that when the axis tilt eventually reduces to 21.8 degrees the Arctic and Antarctic circles will become smaller. Then the winter sunshine will shine on slightly more of the Earth's surface, i.e. the Sun's rays arriving nearly parallel on to the Earth's surface will strike the Arctic and Antarctic circles at the same point on the actual Earth's sphere — not as it relates to the position of the present land masses, but about two degrees of latitude further north of the Equator in the Northern Hemisphere when the Sun is at the solstice position, and of course the opposite situation applies for the Southern Hemisphere. This is because of the more upright tilt of the Earth's axis.

The tropics of Cancer and Capricorn will be closer to the Equator, which will mean that the temperate zones, possibly, will be cooler but none of this will happen in our lifetime.

The lesser tilt of 21.8 degrees of the Earth's axis may also mean less distortion to the atmosphere by the Sun's gravitational attraction on the atmosphere, and this may mean a more placid weather pattern, but other factors could enter into contention.

Changes in the Moon's declination

The Moon as a separate identity may not alter its present five degree inclination of its orbit to the ecliptic so its orbit may not change with the alteration of the Earth's axis to the more perpendicular tilt of 21.8 degrees. So the Moon during its maximum declination period north and south of the Equator would now be about 26.8 degrees (the 21.8 degree axis tilt plus the five degree inclination of the Moon). The Moon would now exceed the Sun's declination during the solstices of summer and winter by five degrees but it would be difficult to anticipate its effect on the weather.

The Sun in its restricted declination of 21.8 degrees north and south of the Equator, may create more heat during the summer in the area between the then tropics of Cancer or Capricorn and the Equator. Some of this heat will be redistributed further north or south of the Equator and into the temperate zones with the assistance of the Moon's extra five degrees of movement to the north and south which will be above that of the Sun's declination.

There must be some speculation on the amount of heat which may be generated and distributed to the temperate zones. In the summer months

Moon at minimum declination of minus 5° from each tilt position

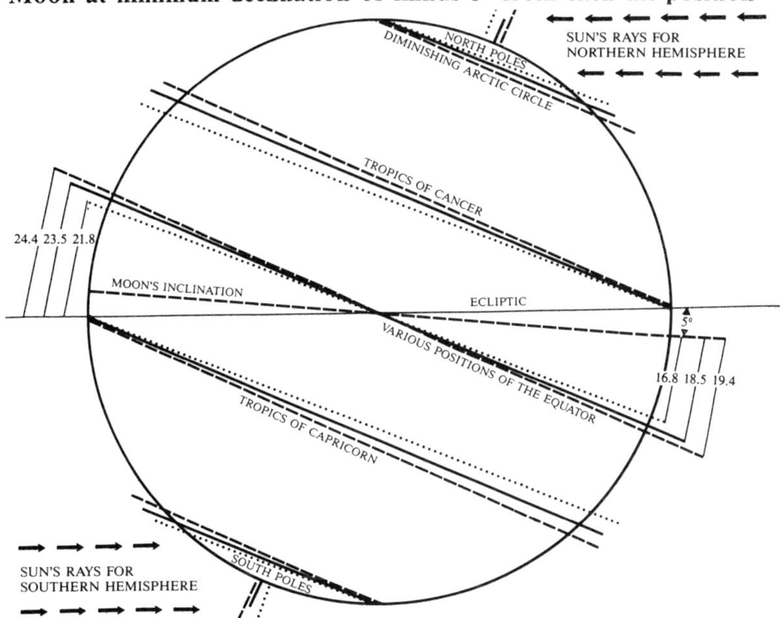

Moon at maximum declination of plus 5° over each tilt position

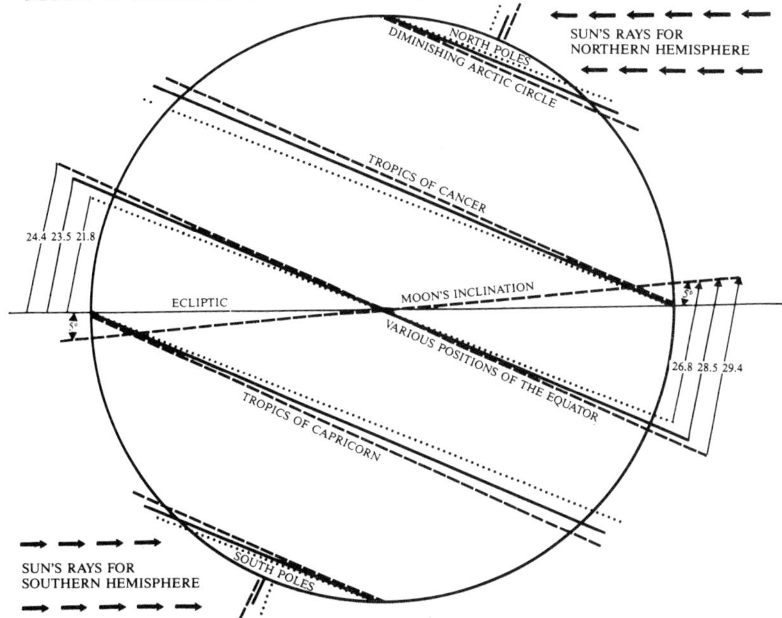

the heat could be considerable within and in the near vicinity of the summer tropic and probably some heat will be transferred into the temperate areas. Because the Sun will shine on more of the surface area near the poles, these areas may also benefit from additional heat, but as the Sun will only be 21.8 degrees away from the Equator at mid-summer, the temperate zones will cool fairly quickly when the Sun begins its return journey towards the Equator and the return to winter conditions may be quite sudden.

However, the time taken for the Sun to move the present 23½ degrees from the tropics to the Equator, will be exactly the same as for the 21.8 degrees minimum declination of the future and this could mean that within the tropics, higher temperatures may last longer there.

In the winter months the colder temperature from the higher latitudes of whatever hemisphere happens to be in the winter season, may be transferred to the latitudes nearer the winter tropic. The overall effect could be colder winters and which would only slowly respond to the summer warmth.

The effect from a minimum declination

After about nine years when the five degree inclination of the Moon is reversed and will slope in the opposite direction to the ecliptic, it will then be subtracted from the Sun's declination. The new minimum declination of the Moon will be about 16.8 degrees north and south of the Equator.

This declination movement of the Moon will be about five degrees inside the Sun's declination of 21.8 degrees north and south of the Equator.

What effect this will have on the world's climate would still be anyone's guess. Possibly, in the summer months, the heat will move away from the tropics and into the temperate zone in a more restricted manner; and because of the lesser declination of the Moon north and south of the Equator, it may also restrict the movement of the atmosphere away from the tropics towards the poles, so the Arctic and Antarctic ice may advance nearer to the tropics.

The habitable parts of the world could be restricted, in part, to those places within 40 degrees latitude north and south of the Equator with the more hardy races, such as the Eskimos, living in the countries near the Arctic Circle as they do at present.

The phases of the Moon may also play an important part in the temperature range each month, especially near the tropics and as at present, the perigee cycle.

The hemisphere in the winter season could be extremely cold and, as mentioned, with the ice and frozen seas advancing towards the tropics, and this may affect latitudes near 50/55 degrees north or south of the Equator. This figure is but a guess, for the effect could be nearer 45/50 degrees, especially in the Northern Hemisphere, where the Baltic Sea does occassionally freeze now, so it may not be impossible for seas to freeze around the British Isles and south of Alaska to the Aleutian Islands. However, before we will know if this projection will be anywhere near correct, only the next 14,000 years or so will be able to tell us.

The suggested climatic changes in the foregoing explanation come from the idea that the 21.8 degree tilt of the Earth's axis will confine most of the heat from the Sun's rays to a narrower latitude range, this being closer to the Equator than it is with its present 23½ degree tilt, and that the areas closer to the new Arctic or Antarctic Circle could be cooler for that reason, especially in the winter season. But as has been mentioned, the Arctic and Antarctic circles will be closer to each of the poles and this in turn means more sunshine two degrees closer to the poles in the winter than it is at present, even if the hours of sunshine are of short duration. In the summer, the area where the Sun remains above the horizon for 24 hours a day will be reduced in size (the Arctic and Antarctic circle) and in the areas which have sunshine now for the 24 hours, the Sun will set for a short while with the changed axis tilt.

Each of these changes could have some effect on the temperatures but in general terms it is felt that the results will be towards negative anomalies rather than being positive especially during the period when the Moon's inclination is subtracted from the Sun's declination which will give a 16.8 degree lunar movement north and south of the Equator, thereby reducing the outflow of heat from the tropics and towards the poles by the restricted atmospheric movement.

Most of the climatic changes which we will experience in our lifetime will be from the changes within the Moon's orbit as it relates to the Earth and the Earth's normal seasonal advancement during the year.

The present astronomical situation will provide the periods of drought, excessive rainfall, the more benign seasons, whether summer, autumn, winter or spring, which are all the result of the Moon's orbital relationship with the Earth. All the various seasonal changes in the weather will continue to differ in most of the countries throughout the world in any one year. One country may have a serious drought, while cyclones lash other countries. Snow storms may blanket large areas of the countryside in gentle falls, while elsewhere blizzards will take their toll in life and destruction. So there cannot be a world-wide climatic condition because different weather trends occur in different countries in the same period.

To use averages to show, say, lower average temperatures throughout the world, only takes out of context the fact that one country will have higher averages, while another will have lower averages. The same two countries could have a reverse effect a year later and still show the same world-wide average as before. Averages will tell one very little.

All countries have good years and bad and some traditionally will have more bad years than others. Those countries with traditionally bad years usually manage to string a number of consecutive bad years together and only an odd good year at any one time.

Large countries or continents may experience about every vagrancy of the weather at the same time by excessive rain in one part, drought in another and good climatic conditions elsewhere.

So how accurately can climate be defined?

Do climatologists understand climate

Climatologists have been writing on this subject since the word 'climate' came into being. They will explain in detail the climate in any given period, what effect the weather had at that time on the community at large, its economic gain or loss, what crops should have been grown, how much was harvested. Yes, every conceiveable detail from the past is written down. They are wise after the event but never before.

When a climatologist does venture into the future, it is always vague, such as — 'There is a cooling or warming trend. We can expect to see dry conditions in the next few years'. Every farmer knows he must expect dry years every now and again.

Or the comment will be: 'we can expect another ice age to develop because the present century is the mildest for the past thousand years'. But never a mention on why a colder spell will develop nor exactly when.

When a volcano blows its top and spews copious amounts of ash into the atmosphere, it is a climatologist's dream come true. Every prediction imaginable will be made, from colder temperatures to more rain around the hemisphere in which the volcano is situated.

The prediction for cooler temperatures may in part be correct, depending on the season of the year.

But more rain would be a doubtful prediction, one cannot be certain that the rain would not have been there anyway. Also, as rain would wash the dust out of the atmosphere, it would seem unlikely that the dust would remain in the atmosphere in any quantity for very long, and for the further reason that dust would not dictate which type of weather system will form. The Moon governs that area of development, so anticyclones would continue to develop and the normal depressions would arrive in due time with or without the dust in the atmosphere.

It is said that when 60 climatologists meet at a conference there will be 60 different theories on climate and its cause, and each and every one of them attempting to outpoint the other, usually by making statements on a subject matter which can never be proved.

Have all encompassing ice ages been factual?

Writing on ice ages offers the climatologists an opportunity to have a field day in their attempts to outdo one another in explaining cause and effect.

None, or possibly only an odd climatologist, have offered a possible moon cause within the astronomical reasons which have been offered for most, if not all of the ice ages, it has been claimed that we have experienced on Earth.

The precession of the poles in which the Earth's axis describes a circle as viewed from space — the circle taking about 26,000 years to complete — only changes the seasons with respect to our calendar. Even with the Moon's changing orbit over the years, the precession of the Earth's axis will have no influence other than changing the dates of the seasons.

The influence of the Moon under those circumstances, and under the present situation (while the Earth's axis tilt is 23½ degrees), may cause seasons to become colder, mostly when the Moon is at the minimum declination of about 18 degrees. The summers may become cooler and the winters cold and such a cool spell could last a few years, but taking the Moon's influence on its own, it would be very doubtful that an ice age could develop, but if the Sun's output was in some way reduced, then a mini ice age could be possible.

The lack of sunspots on the sun's surface has been linked to mini ice ages in the more recent centuries, but without the required astronomical information it is not possible to say whether the Moon was at a declination, or a perigee situation, which when related to a minimum sunspot period, would have coincided to induce a mini ice age.

One mini ice age mentioned as being between 1550 and about 1850 is far too long to suggest a lunar influence, and its distance in years from the present period, a mere 400 years, is too short to suggest an astronomical influence such as a difference in the Earth's orbit around the Sun, or an alteration in the tilt of the Earth's axis.

A decrease in sunspot activity for 300 years also seems to be too long in time.

For volcanic activity to make such an impression would require continuous spewing out of debris into the atmosphere for most of the 300 years, and there have been no reports suggesting that possibility.

So if there were no factors on Earth, or in its orbit, its tilt or a lunar influence, it seems that some influence in space would be about the only other factor to consider.

The other planets are too far away to do more than offer a slight interference to the Earth in its orbit about the Sun, and as for acting gravitationally on the atmosphere and creating such an impression on it that the gravitational attraction will alter our weather, then this suggestion appears to be impossible. Even the alignment of the planets is no solution to the development of a mini ice age. The Earth has just been associated with such a close alignment and a mini ice age has not developed from that cause.

Space debris

For a mini ice age to develop, one of the main considerations needing investigation would have to be a reduction in the amount of sunlight received on Earth. Most of the probable reasons that could reduce sunlight on the Earth have been mentioned, and one of these is debris or dust in the atmosphere, which must be discounted as a reason for the previous mini ice age.

But what about dust in space.

It is known that explosions of one sort or another occur deep in space from time to time.

Our Sun is but part of a spiralling galaxy in the universe, and the Sun is situated in one of the spiralling arms within the 'Milky Way' galaxy. This galaxy, as it is with all galaxies, is continually on the move and the arms of the galaxy are moving around the centre of the galaxy. So, while the arm of the galaxy containing our Sun is moving through space, it could be possible for that arm to move into a huge cloud of debris moving through the universe and being intercepted by our Sun and the associated planets.

The amount of debris may not necessarily be compact as far as space is concerned, but any increase in the number of particles between the Sun and the Earth could be sufficient in volume to reduce the amount of effective sunlight reaching the Earth, and possibly offering a reason for the reduction in sunspot activity which, it is stated, also reduces the temperatures on Earth, especially over the winter months.

A cloud of debris in space could easily take 300 years to drift through our section of the universe, or for our section of the galaxy to move through such a cloud, for there is nothing which will indicate what comes first, the cloud moving towards us, or us towards the cloud.

If our particular part of the solar system did in fact pass through a cosmic cloud of dust, it may not leave much evidence on Earth. The dust could be burnt up in the atmosphere before it had a chance to settle so there would be no dust for analysis to prove that such an occurence did in fact take place. However, some evidence that the Sun had passed through a cosmic cloud could show in the variation on the magnitude or brightness of known stars visible to us from Earth.

A scale indicating the brightness of one star as against another has been in existence for many years, and provided the scale has a solid base from which to judge brightness, then any alteration in brightness would show, and if the drop in brightness lasted for a number of years, then some evidence that there was a cloud of cosmic debris in space could be deduced.

Unfortunately, if the brightness of one star as against another star was judged from the brightest star known, without some method of ensuring that that star's brightness was set against a constant value which was understood here on Earth, then should that brightest star lose some of its magnitude, all other stars would also dim in equal proportions, but the scale of magnitude of one star against another would still remain constant, one to the other.

In the use of the electric light bulb, we use the number of watts as a guide to brightness, as say, 25, 60, 75, 100 watts etc. and this is a known standard.

So would we know if our part of the solar system did in fact pass through a cosmic cloud during any period in the past when a mini ice age affected Earth?

During that particular mini ice age, much was made of the fact that the River Thames froze over many times, and this does suggest a rain-free climate, and further suggests anticyclonic conditions. That factor raises another question as anticyclones are formed from air which is heated in the equatorial regions rising into the upper atmosphere and cooling as it moves towards the temperate zones where it descends to create cells of high pressure. The temperatures near the Equator during a mini ice age may not be particularly high, but sufficient heat may be generated to cause the atmosphere to circulate as a Hadley Cell. The descending air will be considerably colder than that experienced at the present time and sufficiently cold to freeze rivers or even seas. This in turn would seem to indicate that anticyclones will form irrespective of the amount of heat in the atmosphere. There was no mention that the weather was unsettled previously, to form the icy conditions so it has to be presumed that the ice was the result of a cold anticyclonic situation.

Most literature written on mini ice ages is rather vague beyond the centuries between the 1500-1800 years, and even if more was written on this subject, its content would not be any more enlightening than that which is already available. If the conditions were anticyclonic then, what were the barometric pressure readings and how would they compare with today's pressures? It is appreciated there can be no way of obtaining a comparison between then and now but the thought could offer some interesting speculation.

A climatologist's view on climate

Climatologists, from their reported comments, would prefer to concentrate on that area where speculation is the more important ingredient. To talk about tens of thousands of years ago, or into the millions of years is an area which is rich in guesswork. Nothing tangible will come out of it that can possibly assist us with the immediate climate of today.

The variables are immense from what may have occurred then, to what can happen now or in the useful future. Useful in the sense that some organisation by man will be possible to counter droughts, floods, excessive heat or temperatures.

Few climatologists have yet said what the next, say, 10 years will be, whether milder, colder, dryer or wetter or whatever meteorological changes in the weather could take place, although in more recent times some speculation has been offered on what type of conditions may arise from the 'greenhouse' effect and offering an outlook for a reasonably short period ahead under those circumstances.

Some predict that the slowly increasing accumulation of carbon dioxide in the atmosphere will produce the 'greenhouse' effect; that this higher quantity of carbon dioxide in the atmosphere will restrict the re-radiation of the incoming heat rays which will then slowly accumulate on Earth, with the result that global temperatures will rise. The ice in the Arctic and the Antarctic will melt, causing the seas to rise to something in the vicinity of 300 feet in the long term, thereby flooding low lying coastal areas, but more on this subject in another chapter.

A number of climatologists have said many things on what we may expect in the future, but it is always so far into the distant future that no one living today will be alive when the predictions come to pass, if they do.

Then, of course very few of the prophecies follow the same line of thinking.

Other climatologists refute the thought that temperatures will rise, probably because they have climbed on the band wagon for falling temperatures and a new ice age theory.

How climatologists could believe that the Arctic and Antarctic ice will melt because of overall rising temperatures is difficult to understand.

We are having, we are told, one of the mildest periods in many centuries; glaciers have shrunk to an all time low.

Since the Antarctic has been occupied by different countries for a number of years, including wintering over, which means that research stations have been fully manned all the year round, many reports have come through on the temperature range during both the winter and the Antarctic summer. Although an odd summer has been reported as being milder, most summers appear to be about average. In fact it would be an abnormality if there were not some variations between one year and another and in any case, average or milder temperatures still mean that it is very cold in the Antarctic.

During the summer of 1985/86 a Russian vessel was caught in the southern ice because of an early freezing of the seas further north than normal, and an ice breaker was dispatched to free the ship. This was in February, which could be considered high summer because of the accumulative effects of the 24 hours a day of sunshine there. So if this is a mild period in the Earth's history, why then the early freezing of the seas around Antarctica.

In New Zealand the summer had been mild to warm, and New Zealand is not too many thousands of miles north of the area mentioned.

But the main reason for my suggesting that the Arctic or Antarctic ice will not melt is that the atmosphere near the poles is thin. In other words the depth of the atmosphere slopes from the Equator as a wedge to become thinner at the poles. We know space is freezing cold, and with the thin atmosphere at the poles, the cold of space will penetrate to ground level. Coupled with this thin atmosphere, there is the long winter night when the Sun does not shine for about six months, and even when the Sun is above the horizon, possibly for three or four of the months just after rising and just before setting for the long winter, the Sun is only just shining and without much heat. So how can any climatologist justify the claim that the ice at the poles will melt significantly and raise the sea levels through a warming atmosphere.

Even if the radiation output from the Sun was to increase dramatically and heat the atmosphere at the Equator considerably, the long winter's night at both poles would ensure that the ice there would stay frozen.

Longer ice ages: possible causes and effect

Ice ages lasting thousands of years are mentioned by a number of climatologists and theories have been promulgated as to when these ice ages took place and why, or the mechanism which allows the ice age to occur.

In 1920, a Yugoslavian geophysicist named Milankovitch, advanced a theory that the Earth's orbit around the Sun changed over or on about a 100,000 year cycle from very elliptical to circular and back to elliptical.

He was also responsible for the theory on the Earth's tilt altering from more than 24 degrees and down to less than 22 degrees, which theory has been covered elsewhere.

At the present time the Earth's orbit around the Sun is slightly elliptical. This places the Earth closer to the Sun in January for the summer season in the Southern Hemisphere. This position is called perihelion, while aphelion, when the Earth is at its greatest distance from the sun, is in July which is the summer season of the Northern Hemisphere. The difference in distance is approximately three million miles.

When the Earth is at perihelion, the orbital speed of the Earth is faster than when it is at aphelion, which means that the summer of the Southern Hemisphere is shorter, by a few days only, than that of the Northern Hemisphere. It is also warmer.

Conversely, the winter over the Southern Hemisphere is longer, while the winter of the Northern Hemisphere is shorter, which should translate into colder winters in the south and milder in the north, but in actual fact it would be a very difficult point to prove.

Reference could be made to rising or falling average temperatures, but averages only tell us that there were cool summers and warm summers, cold winters and less cold winters. Also summers or winters in each hemisphere do vary in the same summer or winter at different longitudes. This fact has been remarked upon many times.

The effect from an extended orbit

Just how elliptical the Earth's orbit will extend is not known, but if the present distance of three million miles difference between perihelion and aphelion in their radii from the Sun changed to say, 11 million miles, as stated by Milankovitch, and the seasons remained in their present form with the solstices at the extended part of the Earth's orbit, then some very noticeable changes in climate would take place.

This is assuming that the Earth will be equidistant from the Sun to the solstices on one hand and the equinoxes on the other for their sections of the orbit. But it could be possible that the extended orbit may create an extended aphelion placing the Earth further from the Sun and a perihelion with the Earth closer to the Sun than at present. This would be an orbit resembling that of a comet using the Sun's gravity to swing it on its orbital path, but for the purposes of the following explanation, it will be assumed that no aphelion or perihelion exists.

The Earth will orbit faster on the middle section of the narrower part of the elongated orbit and slower through the solstice section of the orbit. The summer of the Southern Hemisphere would change dramatically. Because of this the summer would be extended somewhat because the Earth would take longer to pass through the solstice position but probably cooler because of the greater distance the Earth would be from the Sun. The growing season would be affected. How the rainfall would be affected would be impossible to anticipate. Less heat may reduce evaporation thus less cloud cover to the extent that sunny conditions could predominate, especially towards the higher latitudes.

Any increase in the snow and ice at the pole experiencing the longer winter would naturally reduce the level of the seas and of course increasing the size of the land masses.

The winters over the Southern Hemisphere would be longer and colder with snow and ice advancing a little closer to the tropics than it does at present.

The overall effect on the climate seems likely to be more severe than at present.

Over the Northern Hemisphere, the summers also will be somewhat longer, because the Earth will be orbiting more slowly through that section of the orbit also, and cooler than at present because the Earth will be as far from the Sun as over the southern summer. Whether the longer summer will cancel out the cooler temperatures, only experience would tell us. But ice would undoubtedly remain into the summer in latitudes closer to the tropics than at present.

The winter, like the Southern Hemisphere's winter, would still be cold.

At the moment the Moon's influence on the atmosphere is not being considered in detail.

However other climatic changes will occur because a wobble of the Earth's axis describing a circle as viewed from space over a 26,000 year cycle, will mean that ultimately the summers will change places with the winters. The cycle will also have the summers and winters occurring on the longer side of the ellipse as the seasons alter their position on the Earth's orbit around the Sun.

Because the Earth within its orbit changes the shape of the ellipse over a period of approximately 100,000 years and the wobble cycle is about

26,000 years, it follows that the wobble cycle — that which changes the place of the seasons on the Earth's orbit — will take place about four and a half times during the time the ellipse changes its shape once from near circular to an extended ellipse. This will mean that the Earth will be going through one of its wobble cycles at least once when the Earth's orbit is at its maximum ellipse.

Now, if the summers and winters take place on the narrower diameter section of the ellipse, the summers and the winters will be close to equal length over both hemispheres, and because the Earth will be about equidistant to the Sun at these times, the summers of each hemisphere will be warmer than they are at present and the winters also could be milder than at present, and with no ice age.

The equinoxes, as mentioned, will take place at each of the extended parts of the ellipse respectively. Each of the equinoxes always will be positioned at its own end of the ellipse and each equinox would be cooler and last longer than now, for this would be the time when the orbital speed of the Earth would be at its minimum speed. While approaching the autumn equinox, the effect may be similar to an early start to a winter, for it follows that when the Earth on its orbit arrived at the winter solstice position after the autumn season, the temperatures may already be lower than normal because the Earth's speed will be diminishing by then so the winter temperatures could stay cool and may gradually cool even further for a while.

The Earth continuing on its orbit would begin to pick up speed through the solstice position and as it approaches the spring equinox and because the faster orbital speed will slow down for the spring equinox, the temperatures will be slow to rise until the Earth gathers speed when approaching the summer solstice. Now being reasonably close to the Sun, the summer would be warmer than now for a shorter period also, then cooling more quickly as the Earth moves towards the autumn equinox on the Earth's orbit.

From the above comments it would be unlikely that an ice age would be the result of an extended orbit. The autumn and winter temperatures would not be low enough, or long enough and the spring equinox, while still likely to be cool or cold and then followed by a warm summer, would quickly cancel out the cooler and longer autumn and the quicker winter section.

So it would seem that the combination of the wobble of the Earth's axis and an alteration in the elliptic of the Earth's orbit around the Sun will not produce an all encompassing ice age even although the average temperatures were to be lower than at present.

The circular orbit has not been covered, because it obviously would not create conditions worse than those experienced at present.

If an ice age of the magnitude that climatologists have stated had taken place in the past and lasted thousands of years, some other factor must be the cause.

That leaves the alteration in the Earth's tilt on its axis with a 40,000 year cycle which varies from under 22 degrees to above 24 degrees and it seems unlikely that the maximum tilt of over 24 degrees which would be about one degree above that of 23½ degrees, which it is at present, could cause an increase in the formation of ice, in fact a decrease would be the most likely result because of the extended tropics.

But a decrease in the tilt of the Earth's axis to under 22 degrees and with the Sun's radiation output restricted to a narrower band at the Equator, then such a possibility could exist. The tropics would be only 21.8 degrees north and south of the Equator. Even then the increase in ice could only be in the latitudes at a reasonable distance from the tropics. What effect a 21.8 degree tilt of the Earth's axis would produce while the Earth was going through a period of an extended orbit would be difficult to assess, but it could possibly create a situation whereby ice may lie longer a few degrees of latitude closer to the Equator. But as the tilt cycle is over about 40,000 years and the Earth's orbital change is about 100,000 years, the period it may take for the tilt cycle to occur at the exact period whereby one cycle fitted in with the other cycle so that a longer cooler spell may develop, could be in the order of hundreds of thousands of years and may be in the millions.

At the present time, the winters can be severe and some Northern Hemisphere continents do have snow and ice throughout the winter, but it always melts with the arrival of summer, and much the same circumstances would apply with the minimum axis tilt, except the ice would remain a few degrees of latitude nearer the tropics, and certainly not in sufficient quantity to cause an ice age lasting all those thousands of years, a period of time which is always mentioned as being part of earlier ice ages.

To add in the gravitational influence of the Moon on the atmosphere, to that which has already been written, to the various changes which occur progressively to the Earth's orbit around the Sun, could mean an increase in ice every few thousand years, but to last thousands of years seems to be an exaggeration of the facts.

All that which has been written on the subject of ice ages is but speculation only, we will never know what will be the true position.

The only place on Earth where ice does lie all the year round, and may have done so for hundreds of thousands of years, is at the poles.

Do the poles shift regularly?

Could it be possible that the poles do alter their positions slowly over thousands of years, i.e. as distinct from the continental drift theory when polar land masses could drift away from the poles, so that ice will cover a new area and release other areas from the permanent grip of the ice.

It is known that the polarity of the Earth has changed in the geological past and this change is permanently set into ancient rocks. The normal magnetism in the ancient rocks is also aligned in a direction which does not now equate with the present north-south alignment of the magnetic poles and which is caused mainly by the changing position of the continents over millions of years. But these changes only tell us where the north magnetic poles may have been and it bears no relationship to where the north or south poles may have been in that distant geological past.

At present the north magnetic pole is situated a few degrees away from the North Pole, so it could be that the north magnetic pole has always remained in the same place, but the North Pole may have wandered and changed its position over a period of thousands of years.

As everyone knows, the Earth is near enough to a sphere, being slightly extended at the Equator and flattened at the poles.

The interior of the Earth is more or less plastic or pliable. The Earth also spins on its axis once a day and the speed of rotation at the Equator is about 1609kmh (1000mph). So how is it that the Earth remains a sphere? Such a speed at the Equator should cause the Equator to extend by centrifugal force over a period of time and flatten at the poles, more than it is at present, so that in time the Earth will begin to look like a doughnut.

As mentioned previously, the axis of the Earth describes a circle as viewed from space, the circle completing one circuit in about 26,000 years.

Is the circle of this circuit exact or could it be that by some means the poles may shift their position on Earth by not returning to the same part of the circle which it left at the beginning of the cycle, but say to a point one or two degrees away. If it did not return to its original position, it would suggest that the north and south poles may wander away from their present positions over that 26,000 year period.

The change would be very gradual, but over hundreds of thousands of years, or millions of years, the alteration in the position of the poles would be very positive. Such changes, if they do in fact occur, would also assist in keeping the Earth in a spherical shape.

The position of the Equator would shift too, as one would expect, but wherever the changed position of the Equator was situated, part of the old equator line would still be part of the new equator, although it would be on a different angle. This situation would still apply, even if the poles were to shift 45 degrees over a lengthy period of time, although it is not suggested that such a massive shift would in fact take place.

A shift by the poles in this manner could explain the massive ice ages lasting thousands of years, which are known to have affected large areas of the Earth's surface. The areas affected would always equate with the Arctic and Antarctic in the size they would be at any particular time. There seems to be no proof that all the Earth suffered an ice age at the same time; in fact it would be illogical to assume that such was the position.

If an ice age of those proportions did take place, it would have to entail a complete blocking out of the Sun's rays, and if that happened, where would the moisture come from to cover large parts of the Earth with ice? For there to be any ice on land, it would require that water be frozen, and that water would have to fall from the sky as snow to form the ice, and without heat to evaporate water to form cloud to precipitate as rain or snow, no ice could form on land. The seas and lakes may freeze and the soil may freeze, but without a large and thick coating of ice.

Most of what has been written by various authors on ice ages has not specifically mentioned a complete covering of the Earth with ice, nor has it been stated that it was not so covered. It is by implication only that it could be inferred that ice, at some time in the distant past, did in fact cover much of the Earth at the same time.

The fact that it has not been specifically mentioned one way or the other, whether the Earth was or was not almost covered with ice could be a deliberate ploy to keep both points of view alive and shift from one to the other as expediency determined.

The suggestion that the movement of the north and south poles away from their present positions could be the cause of ice ages, known to have occurred over other parts of the Earth in ancient times, could be termed imaginative, but it is the only explanation which covers the facts. Only mini ice ages for a relatively few years would be possible otherwise, outside of the ice enlarging an area around the poles, which will occur anyway from time to time by the reduction in the Earth's axis tilt and even this event will be many thousands of years apart.

12: Causes of droughts and other phenomena

It is not intended to go into detail on specific droughts, some of which have lasted a number of years, in Africa particularly. In some respects the severity of the droughts there can be blamed on mismanagement of the land by its inhabitants by overgrazing, the removal of trees and the nomadic lifestyle where the people ruin an area and then move on to despoil more ground successively.

The rains may fail, but in some districts the normal rainfall may be sufficient only to sustain growth at a low level anyway, so when that amount of rainfall is reduced, drought conditions must follow.

Outside of these normally dry countries, most of the world suffers from dry spells or larger droughts from time to time, lasting for a season or up to many years.

Countries which benefit from the monsoon rains will occasionally see the monsoons fail and the usually dry season preceding the monsoons will be extended to blend in with the next dry season. Countries such as India have experienced plenty of droughts for this reason.

Sometimes the monsoon failure will be selective and the rains will miss some areas and not others.

In the temperate zones in both hemispheres, droughts will affect countries in some sort of rotation and the droughts may be selective in that the west coasts may suffer and at other times it is the turn for the east coasts.

On other occasions the districts closer to the Equator may become dry, and lands closer to the poles at other times. All this apparently at random.

Sometimes all the depressions will circle the hemispheres closer to the poles and their associated cold fronts will move on to western coasts, depositing rain there. If the fronts are reasonably active, then west coasts nearer the tropics will receive some showers anyway, but east coasts on the other side of high mountain ranges will remain dry or comparatively dry. Any south coasts nearer the South Pole in the Southern Hemisphere will also receive their share of the rain and over the Northern Hemisphere the north coasts.

Conditions such as these may last a few years when the east coasts will receive low rainfall totals, which can be mainly from showers moving up the east coasts which are associated with the leading edges of anticyclones drifting eastwards.

The amount of rain from this source will depend very much on the latitude on which the anticyclones are positioned.

The amount of rain from depressions will vary for the same reason. If too close to the poles, the effect will be less in districts closer to the tropics; but if the depressions are at lower latitudes, then more beneficial rain will be the result, above about 45 degrees latitude or less.

The shortage of useful depressions will determine the amount of dryness there may be, and coupled with a number of anticyclones in succession with only weak fronts separating them, will determine the severity of a drought.

Paths taken by anticyclones

The latitudes over which anticyclones drift will determine over which parts of a country the drought will be the most severe.

Climatologists often refer to the changing patterns where anticyclones forsake their 'normal' paths when circling the Earth, either closer to the Equator or possibly closer to the poles.

The 'normal' paths followed by anticyclones are difficult to define, for there is always some change from time to time when anticyclones drift eastwards further south or north than what may be considered their usual path. Sometimes a succession of anticyclones will pass over latitudes as high as 50/60 degrees or more, north or south of the Equator, and at other times they will be nearer the tropics.

When the anticyclones are at higher latitudes, the east winds on the edges closer to the tropics usually give showers on east coasts nearer the tropics, especially should the barometric pressures be fairly low towards the tropics.

When the anticyclones are situated nearer the tropics, on the lower side nearer the poles, the strong west winds will give showers along west coasts from latitudes of about 40 degrees and higher and also any exposed coasts nearer the poles.

It is usual when anticyclones are at the higher latitudes, the depressions will be above them, probably below the tropics.

This situation will give heavy or heavier rain along east coasts in those districts closer to the tropics.

These are the varying situations which can give rise to abundant rainfall or drought conditions over the differing latitudes and coasts in the various countries.

The problem is, of course, to predict when each of those situations may occur.

It has been mentioned that the Moon, in its orbit around the Earth, has an altering declination north and south of the Equator, which is from about 18 degrees and up to about 28 degrees over a period of about nine years, and a return to 18 degrees over the next nine years.

Throughout this book the Moon's influence on the atmosphere has been paramount, and there seems to be no reason why the Moon's position should not be extended to include the latitude on which anticyclones form or the path they may take, any less than the Moon's effect has been shown to be on cyclones.

To say that the Moon's influence is discriminatory will be illogical, so its influence on anticyclones must be there.

The alteration in the declination of the Moon, its latitude as it appears to us on Earth, and its change upwards from 18 degrees to 28 degrees and down again to 18 degrees is over an 18.6 year cycle. It is a repetitive movement by the Moon as it orbits the Earth on its five degree inclined orbit to the ecliptic whereby it progresses from the north to the south of the Equator and back again, (if one could view the Moon's orbit from far out in space, it would show as orbiting, near enough, from one tropic to the other in a straight line) and it is variable from month to month in that sometimes the steady increase from the 18 degrees in declination will reduce slightly, by minutes of a degree, for two or three months when the Moon will again begin to increase its declination for about four months to repeat the previous action and again reduce slightly. This steady alteration will continue until the 28 degree declination is reached.

So if the Moon does affect the anticyclones and where they form, then that positioning should relate to those degrees of declination the Moon is at during that particular year.

Now, if an anticyclone is over the Southern Hemisphere and the Moon is at its southernmost declination point — and at whatever number of degrees the moon may be at, at that time — the anticyclone will usually drift east on the latitude on which it has developed, but as the Moon moves northwards through the latitude lines and towards the Equator, the anticyclone will drift in a north-easterly direction with that northern movement of the Moon.

The reverse situation will take place when the Moon is going south, an anticyclone will have a tendency to go further south in a south-easterly direction, but only when the Moon has crossed the Equator and arrived over the Southern Hemisphere.

For some reason, when the Moon is over the opposite hemisphere, its effect seems to be lessened in the other hemisphere but the Moon's action on the anticyclones is positive and understandable when it is over whichever hemisphere.

This means that when the Moon is over the Northern Hemisphere, other rules for the Southern Hemisphere, which are not completely understood, apply.

This is a general rule, but occasionally there are exceptions, as there are with most rules.

However, overall, the changing declination of the Moon year by year will also alter the latitudes on which the anticyclones or depressions will be positioned and therefore the climate over the world.

The increasing or decreasing declination of the Moon north and south of the Equator, as mentioned on a number of occasions, is the result of the precession of the five degree tilt, or inclination, of the Moon's orbit as it relates to the ecliptic, the plane of the Earth's orbit around the Sun. This precession is added to or subtracted from the 23½ degree tilt of the Earth's axis. The point where the Moon rises above or drops below the ecliptic is referred to as the 'nodes' and this node position precesses by about 1½ degrees each lunar month, which means that in about 12 months, the nodes position will change, altogether, by about 20 degrees.

As the north and south maximum inclination of the Moon moves away from the solstice position so will the declination alter, as related to the lines of latitude on Earth, by reducing or increasing depending on whether the declination is on the way up or down. This changing node position seems to have an effect on the world's weather and it may be part of the answer for the differing locations on where droughts may occur, in that the droughts move on to new longitudes progessively. The position of the phases of the Moon, which are always changing also, will have their influence on the climate too.

The rising and reducing declination of the Moon each month north and south of the Equator as it orbits the Earth, can be likened to a nut on a bolt being turned up or down on the thread, but because the Moon is constantly moving between each hemisphere each lunar month, the weather systems will not remain at the same latitudes during the month. They will change, depending on where the Moon is positioned at any given time. It is therefore not easy to notice a slow latitude change taking place over the years with the weather systems.

The Sun also will have some influence through its own gravitational attraction on the atmosphere and because the Sun will change its own declination over a hemisphere during the length of a lunar month; so by the time the Moon has returned to its previous declination position the next month, some alteration in the position of the weather systems, or within the systems themselves, will have taken place through the combined gravitational influence of the Sun and Moon and because the Moon phases will have changed also.

Even between one year and the next, the changes which take place astronomically are significant. About the only period when a comparison between one year and the next may be possible, is when the Moon is either at its maximum or minimum declination of either 18 degrees or 28 degrees for at these times, and for about three years, the declination of the Moon is almost stationary and it is in the order of only minutes of a degree.

The other weather system of the temperate zones, the depressions with their associated cold or warm fronts, bring the rain so necessary to promote the growth of trees, grass, the fruits and vegetables required by animals and ourselves to live. Without rain the temperate zones would be deserts with smaller populations and little animal life, if any.

Depressions like anticyclones shift about, sometimes forming a skirtwork at higher latitudes nearer the poles and at other times they will form somewhere below the tropics of each hemisphere as distinct from the tropical cyclone developing nearer the Equator.

Although most depressions forming nearer the tropics will drift away to the south-east in the Southern Hemisphere and to the north-east in the Northern Hemisphere, they will, on occasions, drift directly east and now and again slightly towards the Equator. This is most noticeable when the Moon is at its southern or northern most point in its declination. The depressions will follow the latitude line that the depression is on at that time. Then, as the Moon begins to move towards the Equator, the depression may, at times, have a tendency to drift in that direction.

A depression in the opposite hemisphere from that in which the Moon is positioned, usually will not be so affected.

In New Zealand, which is about 2000 kilometres east of Australia and has a length which extends over approximately 12 degrees of latitude, (near enough to 1400km long), and is situated 34-46 degrees south of the Equator, receives most of its electricity supplies from hydro sources. The water is stored in both natural and man-made lakes in each of the two main islands comprising New Zealand.

The bulk of the lakes in the South Island are situated in the south of that island and in the North Island the lakes and the river, with man-made lakes dotted along its length, are situated above the middle of the North Island.

The electrical power supplies between both islands are connected by power cables so that power can be transferred in either direction when necessary.

The lakes in the South Island are fed by both rain, and melting snows from a high range of mountains along the west coasts of the island.

The lakes are mostly situated east of the ranges where the rainfall is low, and only the depressions which cross the island near the area will affect the catchment area and so assist in keeping the lakes somewhere near full. Otherwise slowly melting snow is the alternative for filling the lakes, and of course, that is almost a once a year occurrence.

The North Island's main natural lake, the largest in New Zealand, is also fed by snow, from high volcanic mountains, when it is available, and there can be a number of years when the snow is in short supply. So this lake has to depend on depressions from the northern tropical areas to receive a boost which will lift the lake level to a useful height.

The depressions passing to the south will assist in some measure in filling the lake, but the associated cold fronts, at times, are not necessarily very active.

Over the years it has been very noticeable that depressions will always pass to the south for a succession of years, and our electrical authority will be concerned about the level of the North Island lake.

Then over other years, the problem in reverse will apply. Most of the depressions will pass about or north of the country, overfilling the North Island lake and requiring water to be spilled over the dams and wasted. This same very large lake also feeds the Waikato River on which there are a number of hydroelectric power stations.

When the depressions are north of the country, it follows that snow does not fall often over the mountains to the south, and the hydro lake levels in the South Island begin to suffer.

This is the normal pattern over the years, but whether it follows a predictable cyclic movement is not known as insufficient data, going back many years, is available for a check to be made on this aspect of boom-bust with water storage.

Further, it is only over more recent years that an increasing number of hydroelectrical generating stations have been built on the Waikato River and the need to know when and where rain will fall and in what years has now become more important for the electrical supply authorities.

The problems related to storage lakes for generating electricity from hydro sources have been mentioned to emphasise the differing paths taken by the depressions, in extreme circumstances, when they are drifting east or south-east either to the north of New Zealand or to the south of the country.

In actual fact the change from northern to southern latitudes, or vice versa, by depressions is exactly the same as that which takes place when anticyclones alter their latitudes through the years.

The various depressions will cross New Zealand, through the years, slowly descending or ascending, so that the depressions will ultimately drift over the country on most latitudes progressively.

As mentioned, New Zealand is a fairly long country and comparatively narrow with a range of mountains close to its west coast. The mountains have high peaks some rising to 2000-3200 metres over the length of the South Island and about 1500-1700 metres in the lower North Island.

In the central North Island three volcanic peaks rise to a maximum of 2797 metres.

In the South Island it is the accumulation of winter snow which is important for the welfare of the hydro lakes for it is from the slowly melting snow that the lake levels are replenished. But as the hydro lakes are situated on the drier east side of the main ranges, direct rainfall from the depressions may be insufficient on their own to maintain the water level required in the lakes.

Any attempt to predict the climate of the future based on the explanation offered will also run into problems, for although there is a lunar cycle when there is some agreement between one factor and another, not all factors repeat themselves exactly, so that over a period of several hundred years the cycle then, would have very little in common with the cycle at the present time. There is also a probability that the alterations in the lunar cycles could explain the reason for some mini ice ages.

When it is known exactly when all mini ice ages did occur, a computer may be able to follow the lunar cycles back in time, and check if there could be a duplication or near duplication on all or most astronomical factors involved, which, could be of assistance to us in the future.

A comment which is frequently made in New Zealand, and similar remarks are probably offered in other countries in opposite hemispheres, is that a particular type of season, be it wet or dry, in one hemisphere will be followed by a similar season in the opposite hemisphere. In this country the English season is the one used for comparison, possibly because of the close association New Zealand has always had with England.

If England has suffered a summer drought for instance, farmers in this country are inclined to expect a summer drought to follow here. There is no basis for such a belief nor any soundly based logic to support that view.

Astronomically the situation is completely different and also, the two countries are on different longitudes. If a comparison were to be made it may be more sensible to use a country in the other hemisphere which is on the same longitude. However, other factors would apply, such as the futility in making a comparison between an island country and a continent where the climates could not be compared anyway.

13: Nuclear explosions and weather

Ever since the testing of atomic energy commenced, there has been universal concern that the explosions will affect the weather immediately and ultimately the climate.

While the testing was conducted in the atmosphere, meteorologists and scientists, during the 50s and 60s, were adamant that both weather and climate were being affected.

Any adverse season was hailed as positive proof that an atomic explosion was the sole cause. Early in 1963 an American meteorologist claimed that the 'rugged' winter in USA and Europe was caused by nuclear testing.

As late as 1981 meteorologists were still commenting on that winter and still blaming the explosions.

There was no suitable authority which was able to dispute the claims; it was a statement which was offered as a fact, but without supporting proof.

Every time a drought occurred or a severe storm and even a larger number of tropical cyclones than usual, all were blamed on the atomic explosions. But the true facts are that the testing in the atmosphere was very local and although there would be a violent disturbance in the atmosphere in the area of the explosion, further away, say 100 miles, its effect would be minimal. Even further away the natural movement of the atmosphere would prevail and in a reasonably short time the atmosphere would return to normal in the vicinity of the testing site.

The effect of the explosion on the world's atmosphere would be the same as walking around a room with a lighted candle.

Fortunately for the world's population, atmospheric testing of the atomic bomb has ceased, but underground testing continues in spite of every attempt to ban that type of testing too, especially the Pacific Ocean on the French island of Mururoa. France seems to claim the right to pollute an area well away from their own metropolitan soil. At least the other testing nations now test within their own countries.

The number of articles written, in the past, on 'The Bomb' and its effect on the weather, and also on their causing earthquakes, would fill a library. Some of the articles offer apparently powerful arguments in favour of what the 'bomb' has done to the weather. Analysis has been introduced comparing pre-atomic years with the post atomic years. One article mentioned that over a period of about 50 years before the 'bomb' there were seven large earthquakes and in a few short years after the 'bomb' there were nine.

Serious flooding was similarly treated by attempting to prove that since testing, the catastrophies from tidal waves, avalanches or major floods were infinitely more numerous.

Even the extra number of tornadoes were also blamed on the 'bomb'.

In more recent times nuclear explosions are only occasionally related to weather changes, probably because they no longer take place in the atmosphere.

Is a nuclear winter possible?

After the atomic bomb was dropped on Nagasaki and Hiroshima during World War II with the horrendous loss of life, the suffering of the survivors since that time, and the complete demolition or vaporisation of the cities, a world awareness has grown of the dangers of nuclear weaponry, and general nuclear radiation on people and animal life.

But a danger which is now on the mind of every person is the likely indiscriminate use of nuclear bombs in any possible conflict between the major world powers.

Again much has been written on the probable outcome from such an engagement, no matter how short lived the encounter might be.

Most of the meteorologists and scientists refer to what they now describe as a 'nuclear winter' as a possible outcome from such a conflict.

They mention that nuclear explosions would cause so much dust in the atmosphere of particularly the Northern Hemisphere — because it would be there that a nuclear war would more than likely be waged — so that the sunlight may not penetrate down to ground level with sufficient warmth to keep temperatures at a level which would allow produce to grow.

The suggestion is that summer temperatures could drop to frost level regularly, killing blossom and not allowing fruit to set on fruit trees or seeds to germinate. It has been claimed that the effects would spread to the Southern Hemisphere also so that no part of the world would escape the 'nuclear winter'.

One claim has stated that sunlight could be blotted out by up to 90 per cent and amongst other meteorological problems that may be likely world-wide, the monsoon rains would fail.

Even if growth could be sustained in the Southern Hemisphere, the suggestion is there would be no market for any produce, be it meat, vegetable or general foodstuff.

More recently, opinions on the effect of a 'nuclear winter' over the Southern Hemisphere have been tempered somewhat and the view now held is that the main problem will not be growth but the loss of markets for any produce grown.

The scare stories on a 'nuclear winter' apparently originated from a group of American scientists and what was said by them has been taken up by atmospheric scientists and meteorologists in many countries, echoing virtually the same words on the subject and adding their own interpretations on what the likely effects would be within their own countries.

For many it was an exercise in being first to be in on the act and to gain headlines for themselves in the local or national newspapers or even world-wide.

A question one may ask of these atmospheric scientists or any other meteorologist who has supported the 'nuclear winter' theory is — how do they expect the atmosphere to react after such a nuclear attack. The nuclear bomb would be so destructive that one bomb would wipe out one city. It took only one small atomic bomb for each of the two cities of Nagasaki and Hiroshima in Japan, and the world knows what destruction took place in both of those cities.

The nuclear weaponry of today will be infinitely more powerful and destructive than those first two atomic bombs exploded in Japan.

So, in the event of an inter-continental exchange, if every large city in the USA was attacked in one initial attack and a corresponding attack on some of the large Russian cities in a one wave attack; would the debris from such an attack be sufficient to cover the Northern Hemisphere in such a way that sunlight would be unable to penetrate the supposed quantity of dust and smoke these scientists envisage would be the result of such an exchange.

A sustained attack from both sides in a conflict would be counter productive, because the chaos that would follow should be sufficient for each side to seek a quick truce. If a nuclear conflict was global, involving all nuclear powers and their world-wide bases, there is simply no basis on which to predict how much debris would be hurled into the atmosphere.

But following on from the suggestions made by the atmospheric scientists that debris and smoke will block out sunlight in the event that a massive strike did take place, how do these scientists expect the atmosphere to behave?

Meteorologists say that the spin of the Earth on its axis coupled with rising air in the equatorial regions and descending in the temperate zones will form the anticyclones by compression. So without heat there can be no ascending or descending air to form the anticyclones which assist in setting the atmosphere in motion, and so allowing dust and smoke to circulate around the Northern Hemisphere.

Without heat there will be no evaporation, so no cloud, and with no cloud, no rain.

If there were to be no circulation by the atmosphere, there can be no depressions either.

So will the debris from a nuclear attack circle a hemisphere? Could there be a 'nuclear winter'?

Atmospheric scientists would have to be the first to admit that their ideal 'model atmosphere' would cease to function if the events they have portrayed were to come to pass, and that the scare stories propounded by them have little substance in fact. In general, the bulk of the debris under these circumstances will rise from the area of the explosion then slowly settle down over virtually the same area.

However, the Earth has a satellite, the Moon, and the gravitational attraction of the Moon on the atmosphere, if such a nuclear war did take place, could see a continuation of the weather systems circulating within the atmosphere. Much would depend on the phase of the Moon and the season of the year in which the conflict took place, for as it has been mentioned elsewhere, there are times when the normal weather systems remain stationary and on other occasions the systems are constantly on the move.

If a large anticyclone were to be over an area of devastation, the chances would be that very little of the debris would travel very far because there would be very little wind within the centre of an anticyclone; but, if on the other hand there was a succession of depressions moving over the area with the consequential unsettled and fast moving atmosphere, then distribution of the debris would be extensive and could spread very quickly over a wide area.

But depressions are usually accompanied by heavy rain, for cloud would still be about at the beginning of any nuclear attack if it were made at the time the depressions were about, so the possibility would be that most of the debris would be washed out of the sky fairly quickly and before any quantity of the dust had a chance to become suspended in the higher atmosphere.

As mentioned, because this planet has a moon, it seems likely that all the weather systems, as we know them, would continue to operate in a way which would be similar to that at present, and not in the extreme manner that has been propounded by the scientists.

But what cannot be predicted is at what intensity will the weather systems move, i.e. fast or slowly.

If there were no hot spots, there would be no fast rising air, there would be no fast moving air (wind) to replace the rising air. The air movement at the outer edge of the anticyclones would not circulate at a fast pace. The isobars would be widespread. Depressions with extremely low air pressure centres would also seem unlikely, in fact they may resemble small anticyclones circulating in the opposite direction, or simply weak fronts between anticyclones absorbing the friction between two anticyclones following one another.

Although an anticyclone will rotate in the same direction in its own hemisphere, the trailing edge of an anticyclone will move in the opposite direction to the leading edge of the following anticyclone, hence the need for a front, but not necessarily with any rain, to absorb the friction between each anticyclone.

Ultimately without heat it seems unlikely there would be any cloud, and therefore no rain or snow, so the only obstruction in allowing sunlight to enter the atmosphere would be the debris created by the nuclear blasts, and that debris would have to be there in vast quantities to cause events to be as proposed by the scientists.

Dust from volcanoes

Volcanoes can spew vast quantities of ash, dust and vapour into the atmosphere.

When St Helens in the USA erupted in 1980, probably millions of tons of ash were ejected into the atmosphere and up into the stratosphere where it circled the Earth several times and, it was stated, with no noticeable effect on the weather.

Krakatoa exploded in 1883 with an estimated 13 cubic miles of ash sent into the atmosphere and this quantity of ash only sent the temperatures down a fraction of a degree, according to the reports at that time, and even this change could have been a normal cyclic alteration.

How many nuclear bombs would it take to send that amount of debris into the atmosphere and how many more would be required to create a situation where the outcome would be the emotively described 'nuclear winter'?

Nobody would wish to see a nuclear war, nor will anyone want to experience any type of nuclear disaster.

At the time this book was being written, Russia has been going through the pain of the Chernobyl nuclear reactor meltdown disaster in one of its nuclear powered electricity generating plants.

The information coming out of Russia on the number of deaths from radiation sickness, the many others requiring bone marrow transplants, the massive resettlement of tens of thousands of people away from the disaster area probably has focused world attention on the enormity of a nuclear disaster and a greater awareness for the need to avoid nuclear conflict at whatever the cost. It may be that the propounders of the 'nuclear winter' theory were active environmentalists and strong lobbyists on all matters anti-nuclear. They do not really understand the true principles involved which are necessary for the circulation of the Earth's atmosphere and once their minds were focused on a 'nuclear winter' theory, these scientists did not analyse all the other probabilities.

Perhaps the 'nuclear winter' theory was what they required so that they could scare as many as it was possible to scare, in the various countries, regardless of whether they were right or wrong. The scientists also knew that there were very few people who were in a position to contradict them.

In the event, the remaining population is more likely to be poisoned off by radiation and killed off by a breakdown in civilisation than frozen by a 'nuclear winter'. Admittedly the result will be the same.

14: Inducing rain

At some stage most countries suffer from the effects of too little or too much rain and of the two, too little rain is the problem most countries dread.

When a drought begins to bite, farmers particularly begin to look at artificial means for inducing rain, and outside of the ritualistic attempts to encourage rain, which will be covered later, most of, or possibly all experiments, involved seeding clouds with dry ice or silver iodide.

The technique requires cloud of the correct type, and of course moving in the right direction.

An aircraft will fly above the cloud and scatter the dry ice or silver iodide on to the top of the cloud and as the crystals fall through the cloud they will act as a nucleus on which the cloud drops accumulate and when they have accumulated sufficiently and become heavy enough, raindrops will form and will fall as rain, hopefully where it is needed most.

During the 60s and 70s, the USA in particular, spent millions of dollars attempting to induce rain to fall where it was required most. But not everyone was happy with this artificially induced rain. Those that wanted rain when the seeding was successful would not complain, but farmers or orchardists further afield who did not receive the rain when it was expected as normal rain, complained that the rain which fell from the clouds before the clouds reached their area, was stolen from them.

Such a charge could not be proved one way or the other. Nor could it be proved that the rain would have fallen in either area anyway with or without the apparent assistance from cloud seeding.

Where cloud seeding is undertaken close to the borders of another country, relationships can become somewhat strained with accusations and counter accusations on who should have the right to any possible rain which may be required desperately by either party.

In more recent times very little has been heard on the subject of cloud seeding, probably because of the moral issues contained therein.

The USA used the technique on hurricanes in an attempt to ease their fury and Mexico complained the experiments were directing the hurricanes towards that country, so these experiments were discontinued.

Although experiments to modify the weather appear logical enough, the success of the experiments can only be on a very limited scale and these would affect smaller local areas only. The gravitational power of the Moon would need to be overcome first of all before true modification could be considered successful, that is if successful is the correct expression, for what may suit one country or farmer may not be entirely desirable for another party.

Weather control and odd crazy claims

Attempts to modify the weather are not so important, but the accompanying illogical statements and claims made by many atmospheric scientists, meteorologists and other individuals are.

There was the statement that Melbourne, Washington and Moscow will exercise control over the Earth's weather.

How will they agree on who wants what, even if it were possible to control the weather?

As early as 1962 a scientist of the National Science Foundation (USA) claimed that one day it would be possible to create an artificial umbrella to shield us against the burning rays of the Sun. The scientist said that clouds were clearly capable of such modification, but international co-operation would be necessary for the experiments to succeed. Would the scientist who made the statement stand up and explain how it could be done. Where will the Moon be while this modification is in progress?

A psychiatrist from London promoting transcendental meditation claimed TM could change the weather. He was at the time on a world-wide tour promoting TM virtues.

An American scientist claimed that, by driving on the right hand side of the road, tornado activity increased. As added proof he said that on a Saturday there were fewer tornadoes because the heavy trucks were off the road.

A well known British meteorologist wrote that a butterfly flapping its wings could set up a turbulence which in time would be the forerunner of a large storm system. Does anyone believe this statement?

In 1971 an article from London said: "By the turn of the century we may be able to control the Earth's climate so as to be able to dial sunshine or rain for wherever we wish." The prospects are mind boggling. The dialling mechanism would run red hot with everyone changing the weather back and forth according to their particular selfish desires. At the time this book is being written it is 1987, 13 years away from that turn of the century — still no dialling for a weather change is available.

An article by an unknown author was extremely honest when he wrote: "For all the technology at his command, man knows little about the weather, is unable to reliably forecast it, and has no hand at all at controlling it."

A French engineer claimed in 1973 that the proposed 'Chunnel' between England and France would change the climate by materially altering the air flows between the two sides of the channel, creating an 'atmospheric syphon', thus changing the air pressures and therefore the weather.

An incredible statement.

Whenever a country or community suffers from a drought of long duration, eventually there will be a call for a day of prayer, especially in Christian countries. It may be something to do, and it may unite the affected community in a common cause but that is about all that can be achieved by such an exercise.

In the natural scheme of things, the weather is astronomically controlled and to pray for a break in the drought is to ask God to change the orbit of the Moon in a way that will change the weather patterns. Any astronomer, or even any scientist, will say that a change in the Moon's orbit is impossible and even if a change did take place, its effect would be far reaching. Problems not yet dreamed of could be the result.

In 1982 part of the South Island of New Zealand was in the grip of a severe drought. A woman had bought a set of drums at a Nairobi auction some 30 years before. These were known as 'Ugandan Wedding Drums'. They were referred to within the family as 'rain drums' — rain was claimed to have been obtained after people made a small tattoo on them as they walked past. They said it was astonishing how often a favourable result was obtained.

To test the possibility that the drums could assist in bringing rain the owner took them to Oamaru, the centre of the drought. After a tattoo on the drums the rain arrived. A coincidence? Who knows? Perhaps the vibrations from the drums altered the Moon's orbit or possibly agitated the clouds, causing the rain to fall.

From Suva, in 1975, came a report that a Fijian had the ability to control the weather. The Waimare tribe of Naitasiri, 30 miles inland, was asked by the Naitasiri Provincial Council to ensure fine weather for a two day fund-raising festival. It rained heavily in the area overnight and into the morning, but the four days of ritual by the weather wizard had the rain stop before the opening ceremony.

Tribal chief, Esirea Vucaga, said he had to avoid contact with water during the ceremony. He could recommence the rain by spitting into the fire by which he kept his vigil.

Did the fire disperse the clouds or was he just plain lucky?

Sikhgguru in 1976 led 1000 of his followers in a rain festival in London. Jagat Singh HI flew in with his 20 piece orchestra to commence a week long rain-making ceremony in suburban Southall, an area heavily populated by Sikhs and other Asians.

He said he was sure it would rain in four days. Unfortunately there was no follow-up article.

Hong Kong in 1977: anti-aircraft guns firing artificial rain shells were successful in triggering rain in the south-east province of Kwangtung in China. It would probably have rained in any case.

In 1978, Eruera Stirling, a Maori elder from Auckland, NZ, sought a particular star in the east. The Maoris called this star 'Parinuttera'. The way it glowed was used to help forecast the weather. A white tint would mean rain, but a red or green hue meant that the drought would continue. Again no follow-up article, but if there was a change in the star's appearance it would have to show an immediate change in the weather conditions and not a change in the longer term to be really useful.

The Moon at night will frequently have a halo when rain is imminent, or sometimes a faint outline of a moonbow (night rainbow) but the rain usually, will follow within hours.

In 1980 in the Zambian province of Mukuni, hundreds of villagers lay down and some slept at a royal family burial ground, beer was poured on to the graves in liberal quantities. Drums beat well into the night with traditional songs invoking ancestral spirits to come to the aid of the tribe. The increasing frenzied ceremony was to invoke rain and within hours the heavens opened and thunder clapped from the lightning.

Why wait so long to organise this ceremony? Would it not have been more sensible to have held the ceremony before the drought was so severe? Possibly the witch doctors knew the rains were close and waited for first signs of rain which would tell them that now was the correct time to bluff the uninitiated.

A professor from the University of Arizona, an expert on clouds, delivered a paper to a physics conference in Auckland, New Zealand, in 1983. The paper was 'Remote Sensing of Cloud Properties and

Climatological Implications'. He stated there was a close connection between clouds and the climate, and he continued by saying that to modify the weather would involve changing the phases of clouds.

He would be right but the one point he was not aware of; if he had but known it, would have told him that changing the phase of the cloud would be an impossible exercise, for the type of cloud formation is associated with the phase of the Moon and if anyone can change the moon phase from say the first quarter to the last quarter in a couple of days, he would be a magician of the first order. For instance cumulus clouds form mostly when there is no moon in the sky, during the morning with a first quarter moon and during the middle of the day with a full moon, and in the late afternoon with a last quarter moon.

Cirrus type clouds are more likely after the Moon rises, especially during the new moon.

The foregoing of course refer to daytime cloud, the cloud at night will have a tendency to be in reverse.

In 1984 China was to seek assistance from foreign scientists to find ways to move rain clouds in the south of the country to the north-west arid zones. The suggestion was: artificially to induce the warm humid air currents and cloud above the Yangtze to drift to the more northern areas and provide the much needed rain there.

There is little doubt that some atmospheric scientists from the west would have taken the opportunity for a paid holiday to China and then offer their advice on how to achieve this impossible dream. When it would become obvious that there was no future in whatever suggestions that were made in attempting to shift the clouds to the north, the most likely excuse for the failure of the project would be, that the distances were too vast for the project to succeed.

But the true reason for the inability of the scientists to succeed with this futile exercise would never be stated and that the simple reason was, it was never possible in the first instance, but none of the atmospheric scientists would ever have admitted publicly to that one.

Dreaming is permissible in any of the sciences.

In a New Zealand newspaper file 100 years ago it referred to an article in the *Sydney Morning Herald* which speculated on the means by which rain might be encouraged by using an aerial torpedo fired by clockwork or electricity and creating concussion in the upper atmosphere.

In 1985 in Nelson, New Zealand, a pip fruit and kiwifruit orchardist tested an anti-hail device which was designed to set off explosions aimed at disturbing the formation of the ice molecules which produce hail.

The shock waves were effective at an altitude of 8000-10,000 metres where the hail formed. It just goes to show that sometimes there is very little that is new under the Sun.

Claims made by a Mrs Dorothy Mundey of England that she broke the drought in Australia about 1974, are very interesting for Mrs Mundey seemed to have had a particular hatred against cricketers. In 1968 she claimed to have washed out an Australian/England test at Lords. Tennis at Wimbledon was also affected. She said she had nothing against tennis but she was unable to localise the downpour.

Mrs Mundey said she would rain out a New South Wales match against the MCC. When it was mentioned that it was not a test match, her reply was "It doesn't matter, it's all the same to me." Unfortunately there was no follow up story. Her attitude was that the Australians owed her money for breaking a drought.

In 1977 when rain again dogged the Australian cricket tour of England, Mrs Mundey jumped in onto the bandwagon and claimed responsibility. She said she would not stop the rain until the Australians recognised her powers.

She described herself as a hypnotherapist and psychotherapist, and claimed to have ended droughts in other places, such as India, Rhodesia (now Zimbabwe), America and China, but Australia was the one country from which Mrs Mundey demanded payment for her services.

Mrs Mundey claimed she controlled the rain by extra sensory perception and it worked 90 per cent of the time.

There were many other problems Mrs Mundey claimed to be able to cure, such as blindness. We would live in a nightmare if 10 per cent of the population had these supposed powers.

Every now and again extravagant claims have been made on the subject of wars being fought by weather.

As far back as 1968 a top USA scientist stated that within 10 to 20 years, global wars could be fought by using hurricanes, rain and electricity. In 1987 scientists are still attempting to understand how hurricanes develop, and as for altering their paths, meteorologists have enough trouble in assessing which path a hurricane will follow, let alone placing it on a predetermined path of destruction.

On one occasion it was suggested that the USA would manipulate the weather so Communist countries would fall under a new ice age.

Not many years ago the USA had one of the most severe winters recorded, the winter of 1976-77. Now if it was possible to create an ice age, it also should be possible to prevent a severe winter over one's own country. It didn't. Why not?

The USA, Russia and 26 other nations in 1977 were to pledge not to use weather as a weapon in warfare. This seems to be a case of each country bluffing the other that it has advanced far along the road in solving the problem of weather modification, when the truth is that even today no country can say it understands weather completely, let alone controlling weather, be it for warfare or peace.

In 1980 Russia accused the USA of diverting hurricanes from its own south-east coasts and away from Mexico, which in turn was the cause of droughts in that country. Was Russia looking for an admission that the USA did or did not possess such technology?

Some years ago a church group approached me, in my capacity as a long range weather forecaster, to enquire what the weather would be like at the time that had already been scheduled for holding their outdoor religious festival. My reply was that the weather seemed likely to be unsettled.

When the festival was held, the weather was fine and clear and afterwards, in a newspaper release, the church publicly said that they prayed as a group for fine weather and God granted them their wish.

There was never a suggestion that my prognostications were wrong and that the weather was destined to be fine in any case. The best publicity for them, was to prove that the church was all powerful and God's divine will was to shine on their church and their church only.

When the Queen of England was visiting Kiribati, in the South Pacific, for the first time, a tropical downpour was threatening her welcome. Isiabata, well known for his spells in stopping rain, disappeared to make them work. Ten minutes before the Queen stepped on shore, the rain disappeared.

In India before the funeral of the former ruler of Sikkim it had been overcast, cold and with rain, but on the day of the funeral it was bright and mostly cloudless.

A Buddhist monk, Chu Thing was paid $6 to arrange the sunshine through meditation and prayer.

If the powers of Chu Thing are as claimed, many would be prepared to pay, for him, a fortune for his services. But no doubt the reply to such a suggestion would be that as a man of God he could not use his powers for personal material gain. Could the $6 payment be considered personal gain?

15: *Weather forecasts, long range*

In many books written by meteorologists and climatologists, but especially meteorologists, the subject of long range weather forecasts will certainly be mentioned.

Although a number of countries issue a long range forecast, not much is published on their accuracy. One country, England, after many years, finally discontinued compiling their long range weather forecasts. The system used by them was partly analogue after a computer search through past weather records for comparable weather equating with the present period and with the hope that the pattern leading up to the forecast period would continue on into the next period. Such a system was doomed for failure from its commencement.

The USA uses a system involving the charting of the upper air where large scale and slow evolving systems could be related to lower level weather systems. Forecasts were issued for a 30 day period but appear to be generalised and showing dry or wetter spots, warmer or cooler. The forecasts were re-assessed and upgraded after two weeks, or about half the original forecast period. Although it is not known for sure, another two weeks would probably be added at this stage taking the forecast period back to the original 30 days, then repeating the process fortnightly. The need for upgrading after two weeks would suggest that the forecasts were not particularly useful because if an alteration in the forecast was necessary in the second half of the period, then any planning by a farmer or whoever that may have been made for the whole 30 days would have to be re-assessed again in the light of those changes and these could be costly to the user.

Russia follows a system based on synoptic charts, and where some defined areas were affected by a series of low pressure periods or anticyclones, these were then related to a mean track that the depressions or anticyclones may be likely to follow. The weather forecasts cover up to a season ahead. Just how accurate they have been over the years has never been mentioned.

In Germany long range weather forecasts were projected from the general circulation of the air masses over the Atlantic and then on to Europe.

This system relies on the continued movement of the air mass in the same direction and without pausing. So any change in direction or any pause caused by the so called 'block' would immediately put the long range forecast out on a limb.

In New Zealand, commencing about three years before this book was begun, the Meteorological Service was compiling a forecast based on a similar method of projection and for a period covering four days ahead only.

The longer range weather forecast was relayed to the public via radio broadcasts at midday, (since moved to 4pm) on the day previous to the four days of the forecast period. Occasionally guesses are published on what the probabilities may be for the month ahead and now and again the future season.

The main reason why long range forecasts are not completely successful is that the meteorologists do not know what they are looking for, especially if the forecasts need to be specific and in reasonable detail, such as when will it rain or snow, when will it freeze and, in the summer, how long it will be dry or even wet.

Analoguing is really the only method, but this requires a starting point, or a why and when, and also a sound reason for adopting the line being followed.

From what has been written up to this point, it must be obvious that the only way to achieve this objective will be to study the gravitation attraction of the Sun and Moon and its effect on the atmosphere and ultimately the weather.

The author has had a continuing interest in this subject for a period of about 25 years at the time of writing.

In the first instance it was but an idea and based on the fact that there had to be an atmospheric tide.

Over the years, by trial and error, a system has been developed which gives reasonable, and at times excellent results.

There is no A to Z answer in compiling a long range weather forecast, for if there was a straight path it would have been discovered long ago.

It is like a jigsaw puzzle; the pieces must be found and fitted into their correct places where possible. Unfortunately it is only after the forecast period has passed by that one will know if the prognosis was correct. If it should be wrong, all one can do is learn from the experience.

The long range weather forecasts compiled by me are not generalised, but they give as much detailed information as it is possible to offer. The information includes the probable number of days with recordable rain and the likely total rainfall for the month. The month is broken down into heavy rain and shower dates and also the days which are expected to be fine or nearer fine. A temperature range is offered and with a comment on the likely average temperatues and also an indication on the amount of sunshine for the month by stating if below average, average, or above average. General comments are also offered on what the probable growing or other conditions may be like during the month, all of this information is offered for up to 30 districts in New Zealand.

Before this information has been compiled into a long range weather forecast, weather maps are drawn for each day showing the expected position of the weather systems. The relative weather map is selected from a range of weather maps held by me, which go back daily to 1941 and through to the present time.

The selection is by analogue but it is done by using a predetermined formula.

There seems little doubt that my method could be improved upon and become the forerunner for a world-wide system for long range weather forecasts.

Just how far ahead it will be possible to take the forecast will depend on extensive research into the system. Although the principle involved is understood, accuracy can be obtained only by being able to say with certainty why a particular weather map from the past is the correct one to use for a particular day in the future. This requires specialised knowledge, mostly in the field of astronomy which must also be used in conjunction with a knowledge of atmospheric physics.

At the moment my long range forecasts are taken to about 12 months in advance, and in my experience, most users are satisfied with an outlook that far ahead.

However, advantages in being able to take the forecasts further ahead are recognised, both at a farming and horticultural level and into any other projects which are undertaken outdoors, also a truly long range weather forecast must also be useful at government level.

From a farmer's point of view, he will have the knowledge enabling him to alter stock numbers, up or down, in plenty of time for either good years or drought years. If the next winter was interpreted as likely to be severe, he can then build up stocks of hay or grow a suitable crop during the summer to assist in extending the winter feed.

Should heavy snow be in the offing, plenty of time will be available to shift stock to safer pastures and to have the feed on hand at that time.

In the event that a late snowfall was a possibility, the shearing dates for sheep could be altered and so avoid stock losses by having them freeze to death or the lambing dates changed to suit better weather conditions.

The list of possibilities is endless, but one thing is certain; neither meteorologists nor scientists will be able to change the weather to suit any individual farmer's needs, nor would such an ability be desirable.

Long range forecasts using the planets

In every country of the world there will be someone, or a number of people, who compile long range weather forecasts. All will have a formula secret to themselves, but most of the formulae used seem to involve astrology in some form or other.

Why astrology should be so important in weather forecasting, it is difficult to understand. Reduced to its most simple form, by using the distant planets it should be possible to show how they affect our day to day weather and also it should be possible to compile a forecast for the present day; i.e. today, if astrology is a viable proposition, for it is the day to day weather of some previous period that must be projected ahead to become part of the long range forecast compiled for that future use — provided of course that past weather records are used for a projection into the future when using the planets for the system.

Generalised seasonal weather forecasts offering little more information than it will be dry or wet, cold or warm cannot lend themselves to a breakdown which will say what tomorrow, or next week's weather will be like.

Part of this thinking may have been inspired by the theory that sunspots affect our weather and the average sunspot cycle of about 11 years almost coincides with the planet Jupiter's orbit around the Sun which takes 11.86 years.

So far, neither astronomers nor climatologists have brought forward a map of the solar system which shows just where Jupiter was situated on its orbit around the Sun when the sunspots appeared, or on any of Jupiter's successive orbits. The 11.86 years it takes Jupiter to orbit the Sun will still be 11.86 years from whichever point one cares to commence monitoring the orbit. So the question is: does the sunspot cycle precess or advance on the orbit of Jupiter? It is not enough just to relate sunspots to the planet

Jupiter and let it go at that, there must be a sound reason why the sunspots do occur at the time that they appear.

There are eight other planets orbiting the Sun and about the only other massive planet likely to have a positive effect in conjunction with Jupiter would be Saturn, the next planet outside the orbit of Jupiter, and although smaller than Jupiter, Saturn is still the second largest planet orbiting the Sun by a considerable amount.

The mass of Jupiter related to the Earth is 317.8 times as large and Saturn 95.16. Saturn takes 29.46 years to orbit the Sun, which means that Jupiter orbits the Sun near enough to 2½ times to one orbit of Saturn. In the one orbit of Jupiter, Saturn will have advanced about 158 degrees on its own orbit.

To overtake the 158 degrees Saturn has moved forward will take Jupiter, in its next orbit, about another six years or so. The total time it will take Jupiter to catch up to Saturn, assuming each was in the same starting position in relation to one another at first, will be in the vicinity of 18 years which is more than the time taken for one orbit by Jupiter. So this, of course, will not be the same original starting position on the orbit of Saturn. This 18 years is too long a period to equate with the sunspot cycle, but half of that period at nine years is closer. These figures may not be entirely correct and they could be closer to the time taken by Jupiter for one orbit and also the accepted period for the sunspot cycle.

The sunspot cycle is only approximately 11 years and because the cycle rises to a peak before reducing to a minimum, the half orbit between Jupiter and Saturn, when they are together on the same side and when they are 180 degrees apart, could be the catalyst to commence a sunspot cycle.

In other words, the sunspot cycle could commence when the planets are close together on one side of the Sun in their orbits around it and then again when they are about opposite one another and about 180 degrees apart on their orbits, with the Sun more or less in the centre of each of their orbits.

Then it must be remembered that the Earth will continue to orbit the Sun once each year while the planets are so positioned, which means that the Earth will be placed on varying parts within its own orbit, and this means being at differing angles in relationship to each of the planets. At the same time, the said planets will not have moved very far on their own orbits around the Sun.

If the above offers a reason for the sunspot cycles, it still does not answer why.

A well-known Australian long range weather forecaster, Inigo Jones through 'The Long Range Weather Forecasting Trust' which was based in

Sydney, Australia, issued or sold a booklet somewhere about 1949 in which he gave his theories on long range weather forecasting by astrology. This booklet published weather forecasts for Australia, in brief, covering the years 1949-2008.

Inigo Jones also gave his version on when sunspots appeared or disappeared. He referred to the forward movement of the Sun through space in the direction he referred to as 'north' which was towards the star Vega.

When the planet Jupiter or Saturn or both are on the Vega side of the Sun, he stated they act as a shield to the Sun from the magnetic field through which the solar system moves, and he believed that this was the time when there were no sunspots visible on the Sun. He also stated this was the time for droughts in Australia at least, and when the planets were following behind the Sun towards 'north', then this was the time for sunspots and with heavy rain or floods. However, it must be remembered that these two planets would be together on the 'north' or 'south' side of the Sun only rarely, but Jones made use of each of or both planets individually or together for his long range weather forecasts.

It seems hard to imagine that only Australia was affected in this manner by droughts. Surely other countries were receiving normal rain or even floods while Australia was in the grip of a drought, or vice versa. If this was the position world-wide, how can planets be used to say what the weather will be in any given country. The Earth is continually spinning on its axis each day, year in and year out, so every part of the Earth should be affected equally after making allowance for the different season of the year.

Another factor which seems to be overlooked is that the Earth, when orbiting the Sun, adds or subtracts millions of miles between the Earth and whatever planets are supposed to be influencing our weather. Saturn at its closest approach to the Earth, i.e. when the Earth and Saturn are on the same side of the Sun, will be about 1271 million km (790 million miles) distant and at its most distant point about 1593 million km (990 million miles) away when the Earth is on the opposite side of the Sun.

Jupiter at its closest approach is about 627 million km (390 million miles) distant, and 925 million km (575 million miles) at its maximum.

It would be exceedingly doubtful if either or even both planets together would influence our atmosphere by gravitational attraction and certainly not our weather.

Inigo Jones' booklet is very interesting and plausible, but unless his theories could be incorporated into day to day forecasting, it would have very little value for a truly detailed long range weather forecast. There could be some sustenance for seasonal generalisations but it is still not understood in what manner, and when all is said and done the weather of the future must come down to what will the weather be like tomorrow or next week, for when those days arrive, weather maps will have to be drawn to enable an analysis to be done and so produce the written forecast.

If all the long range weather forecasters could be encouraged to publish their methods so that they could be analysed, some positive points may be pinpointed and researched with a view in attempting to understand why that section of a forecast worked or was correct. Or if one particular system was better than another, why was that so.

The problem with any advanced weather forecast, or even a forecast for the next day, is that the forecaster cannot be certain that his forecast will be correct. Sometimes percentages of accuracy are bandied around by meteorological people and by sometimes stating that they are about 85 per cent accurate. The 85 per cent accuracy is OK if you can say also where the 15 per cent inaccuracies will be. Obviously the forecasters cannot say when the inaccuracies will creep in, for if they knew the answer to that one the forecast would be 100 per cent accurate.

Long range weather forecasts can be written in an ambiguous way, so that the inaccuracies will be disguised as much as possible.

But the problem remains the same, for if a long range forecast was as accurate as 75 per cent, or whatever figure one may come up with, it still is the inaccurate percentage where the trouble will lie. Rain may be required and the rain is not there, or sunshine for haymaking or harvesting or just for holidays. The inaccurate percentages will not necessarily be together, they may be spread throughout a month with an odd day or two here and there.

16: Planets — a prerequisite for life

The subject dealing with the possibility that life of some kind may exist on any other planet anywhere in the universe may appear out of place in a book such as this. But the subject does have a place here, because before life can exist, any possible candidate to promulgate any form of life must have weather, and it is in this area some speculation will be offered.

Our planet may be unique in that it hosts intelligent life and many other forms of life such as sea creatures, animals, birds and insects. On Earth we also have the ingredients to support human life, whether it be the flesh of animals, fish, birds, fruit, grain and vegetables. The earth provides grass, leaves, seeds and other products to support the animal life, and in the seas the many small to large mammals prey on the fish and other life forms, while larger fish feed on their smaller kind for their existence and so on down the food chain.

Since the creation of the Earth and the slow development of the early primitive life forms, there were probably as many or more life forms that have become extinct, as there are on Earth at present.

There have been many imaginative stories written on how life began and under what conditions it began. All of these are mainly speculation, for we cannot be sure what the Earth was like in those far off days.

From our present knowledge we do know that there would have to have been weather in some form or another to create the right conditions. Heat and humidity, for instance, would probably have been needed to encourage the development of those early life forms.

Another prerequisite would have been a planet of the right size, at the correct distance from a sun. The sun would also have had to be the correct size, with the right output of heat.

Sufficient water covering parts of the planet would have been needed for evaporation to take place to form clouds, to assist in keeping the temperatures at an acceptable level, and to distribute the water as rain over the drier land masses, so that trees and grass would grow to feed this developing life.

The atmosphere must be at the right depth, if too thick there would be too much pressure, which may restrict the development of life as we know it to be now. If too thin there would be too much harmful radiation reaching the ground, and even the fish could suffer.

The thin atmosphere with its lower pressures could have altered the development of life forms and made them different from those we know at present.

There has to be an ozone layer in the upper atmosphere to screen out or break down the ultra-violet rays, and even higher above the ozone layer, a magnetic field similar to the Van Allen Belt, named after its discoverer, which shields us on Earth against the potentially damaging solar wind.

On Earth we have the four seasons which allow us to grow and harvest cereal crops, grow and ripen fruits etc. for the vast amount of food the world's population needs for sustenance. These seasons are the consequences of the tilt of the Earth on its axis which at present is 23½ degrees and a similar tilt would be required by any other possible developing planet. If this tilt did not exist, the Sun would always shine directly over an Equator and any planet supporting similar life to that on Earth would have only a narrow band for its population to live within, provided it was not too hot there. Without a tilt on its axis, a sun would not have an apparent north or south movement between what we refer to as the tropics.

If the tilt was too excessive, say up to 45 degrees, the seasons would be too extreme and the ice at the poles would probably melt sufficiently during the summer and cause flooding from the higher sea levels, along lower lying coastal areas. Although the winters would cause freezing again at the poles, precipitation, otherwise, may be insufficient, during a summer season, to allow the ice caps to reform to the probable previous levels.

The orbit of a planet around its host sun would have to be as near circular as possible, so that the seasons would be reasonably static year in and year out.

If the orbit was too elliptical, then the seasons would be too variable, and possibly create food growing problems.

The length of the day, the time taken for a planet to complete a revolution on its axis, may be equally important, for if the days were too short, insufficient sunshine during the daylight hours may not allow enough time for the proper development of crops, fruit etc, and if the days were too long, the increased heat may destroy or inhibit growth. This point is purely speculative, for at present, on Earth, the summer days in high latitudes are already long compared to other areas closer to the Equator.

And last of all, a planet must have a satellite which we call 'moon'. This satellite would have to be large enough to offer sufficient gravitational attraction, which is necessary to break up the cloud in the atmosphere. Without a moon, cloud would form by the evaporation of water, from the heat of the sun, on the day side of the planet and that cloud would stay and follow the Sun so that the planet would not receive any direct sunshine at ground level. Under these conditions the days, probably, would be warm and humid, except in the higher latitudes, and the nights generally cold, because the night sky would be clear and because the planet would be spinning on its axis inside the cloud cover facing the Sun.

If the satellite was too small, the cloud may not break up sufficiently, and if it was too large, violent storms could predominate and also the satellite's orbit would need to be, near enough, to the plane of the ecliptic and at a useful distance from the planet. If too far away, it will not have the desired effect on the atmosphere and if too close, its effect could be too destructive.

Without a moon, photosynthesis would be unlikely so nothing would grow, at least not in the manner we understand it to be at present.

Therefore life as we know it would be impossible.

Astronomers often speculate on the possible number of planets there may be in the universe which could support intelligent life, or even life in any form, however primitive or advanced. Perhaps the answer lies in this chapter.

Probably far less than the numbers often bandied about, a number frequently mentioned as being in the millions.

It may well be that we (in our life form) are alone in the universe because of the many factors which are required to sustain life and for it to develop and survive.

Other primitive or totally different forms of life may exist elsewhere in the universe under circumstances which are less favourable than here on Earth.

However, it must be acknowledged that although we have developed to our intellectual level, especially over the past few hundred years, if there is other life in existence 'out there', that life could be as intelligent or even more so than ourselves and it may be that they have enough intelligence to care for their planet in a more caring manner than we at present are managing to do on this planet.

Life on Earth in ancient times

Every now and again palaeontologists make public statements on some important fossil finds of animals and giant birds which had lived on Earth many millions of years ago. Recent fossil finds in Late Cretaceous period rocks (65 million years old) in the inland Hawke's Bay region of New Zealand have been romantically described as a 'monster's graveyard'. The fossils cover a wide range of marine reptiles and the first find in New Zealand of a flying pterosaur. Although whole bodies have not been isolated, one of the more dramatic finds is the head of a mosasaur, a giant lizard-like reptile (pictured overleaf).

In 1975 it was reported from Texas that the fossilised remains of the largest flying creature on Earth had been found. It had a wing span of 15.25 metres (50 feet), which was twice the span of any previously known pterosaur. The fossils were from a period of about 60 million years ago.

In 1980 a fossil of a large bird was found with bones which resembled an Andean Condor, the largest flying bird known today. The fossils indicated a bird with a wing span of 7.6 metres (25 feet) and may have weighed between 72-77 kilograms and it was 1.8 metres (six feet) tall.

The largest of the dinosaurs were apparently about six metres (20 feet) tall and 15.25 metres (50 feet) long, which may have made them the largest animals to roam the Earth in that period.

Most or all of these large animals and birds lived, apparently, from about 200 million years ago and up to about 65 million years ago.

Why they became extinct no one seems to know for sure, although many theories have been propounded and a number of statements have been made that this or that may have happened.

One theory suggested that the dinosaurs in particular may have become extinct through sunburn.

Scientists have commented that the Earth's ozone layer may have been damaged by a change in the magnetic field. It is stated that the magnetic field changes its direction once in about 200,000 years, and during this change in direction the magnetic field is weakened, allowing cosmic rays to penetrate further into the atmosphere. Scientists state further that the nitrogen molecules are split, which in turn interact on the oxygen to make nitric acid and this assists in destroying the ozone layer.

The ozone layer normally absorbed the ultra-violet radiation from the Sun and without this protective ozone layer the more harmful radiation possibly did affect life or even growth habits in the food chain, thus assisting in the possible extinction of the larger animals or birds. This comment is difficult to understand should this be true, for why were all or most forms of life not destroyed also.

The skull and jaws of a fossil mosasaur (a giant, extinct, lizard-like marine reptile) found in Late Cretaceous period rocks in inland Hawke's Bay. Collected and prepared by Joan Wiffen and part of the NZ Geological Survey collection. *Photo by Wendy St. George. NZ Geological Survey.*

Whether this theory can be substantiated or not, we will never know.

Climatologists will advance another theory, that the Earth became too cold during an extensive ice age (how extensive?) and that the large animals unable to acclimatise to the extreme cold became extinct for that reason. They will draw attention to the bison, buffalo or mastodons frozen to death in the Arctic region (where the Arctic is located at present) while still standing, some with grass in their mouths. No reason has been offered for this very sudden freezing of the animals except one that is outside the scope of this book, and controversial. That explanation will not be repeated here.

But the one very important question that needs answering is — how did the large birds of that period manage to fly?

In more recent times the world is familiar with ostriches, emus and the rhea, and also, now extinct, the moa of New Zealand. These birds are all large and flightless. It cannot be assumed the birds became flightless just because no animal predators preyed on them. In fact, in some countries there were predators and to escape, flying would have been the natural way to avoid being caught. But the modern bird did become flightless and yet the larger birds of 65 million years or more ago apparently did fly.

Was the Earth smaller in ancient times?

Could the answer be in the make-up of the atmosphere some 200 million years ago.

Scientists and climatologists seem to have assumed that the composition of the atmosphere was more or less as it is now and that the Earth, other than the drifting of continents, was also much the same as it is at present.

What if the Earth was a good deal smaller than at present and that it has been gradually expanding for whatever reason, possibly from changes in the force of gravity on Earth or from external forces as yet not explained or understood.

If the Earth was smaller and the volume of the atmosphere much the same as now, then it would have been deeper or more compressed, and consequently would have had a considerably higher barometric air pressure at sea or ground level. That atmosphere today, is spread more thinly around the Earth and with a lower air pressure.

The higher air pressure would surely allow the larger birds and bats to fly more easily. Their wings would have had something more substantial to flap upon.

Today most birds are smaller and with light bodies compared to their wing area.

The few flightless birds around now have large bodies and would require a tremendous wing span to enable them to fly. This factor, under the present atmospheric circumstances, would create difficulties in getting the wings to flap hard enough to get off the ground. Also the tendons and muscles would have further difficulty in gaining the purchase necessary to make a long wing flap.

If the air pressure did reduce slowly over the millions of years, as the Earth expanded, it may explain why these birds became flightless or unable to fly or why the less efficient flyers became extinct.

The animal life then, in most instances, was also large and probably sluggish in movement for the reason that the air pressures were higher. The greater air pressure required the evolution of animals that were large boned and solid in stature. When the air pressure decreased, the bodies of dinosaurs in particular did not adapt to the changing air pressure and so they became extinct.

Some of the animals, such as the hippopotamus and alligators survived, but to offset the changes in the air pressure, spent more time in the water as they adapted to the changing atmospheric pressure.

It may well be that seals, sea lions, porpoises and whales became victims of the changing air pressures and switched from land-based animals to become residents of the sea.

Some years ago a paleontologist discovered the skull and teeth of a whale close to Khyber Pass in Pakistan and he came to the conclusion that in the distant past whales had legs and did walk on land. Did the whales take to the sea full time for the very same reason?

However, the drifting continents thrusting upwards into Asia could easily have carried the skeleton of a whale into the area now known as Pakistan and the whale may not have walked there as previously believed.

Was the Moon captured?

Although the suggestion has been made that a slow increase in the diameter of the Earth may have contributed to the extinction or changed lifestyle of the large animal of prehistoric times, the probable change in the size of the Earth may have been more sudden than indicated.

There has often been speculation on when the Earth may have captured the Moon, or if in fact it ever did capture the Moon in the first instance. If not, then it must be acknowledged that the Moon has always been a companion of the Earth since the creation of this part of the universe.

If the Moon was captured at some stage in the distant past, the question is, when and what was its effect on the Earth at that time. For without a moon, the force of gravity may have been fairly strong and creating the need for heavy-boned animals. Once the Moon was captured, then the upheaval and seismic activity and subsequent damage which would follow, and the possible loss in the force of gravity, could have been sufficient to cause the demise of the dinosaurs and the other large animal life of that era.

This suggestion, of course, contradicts a previous comment that life may be impossible on Earth if it was without a moon.

Yes, it is speculation but what part of that distant past is free from that type of speculation?

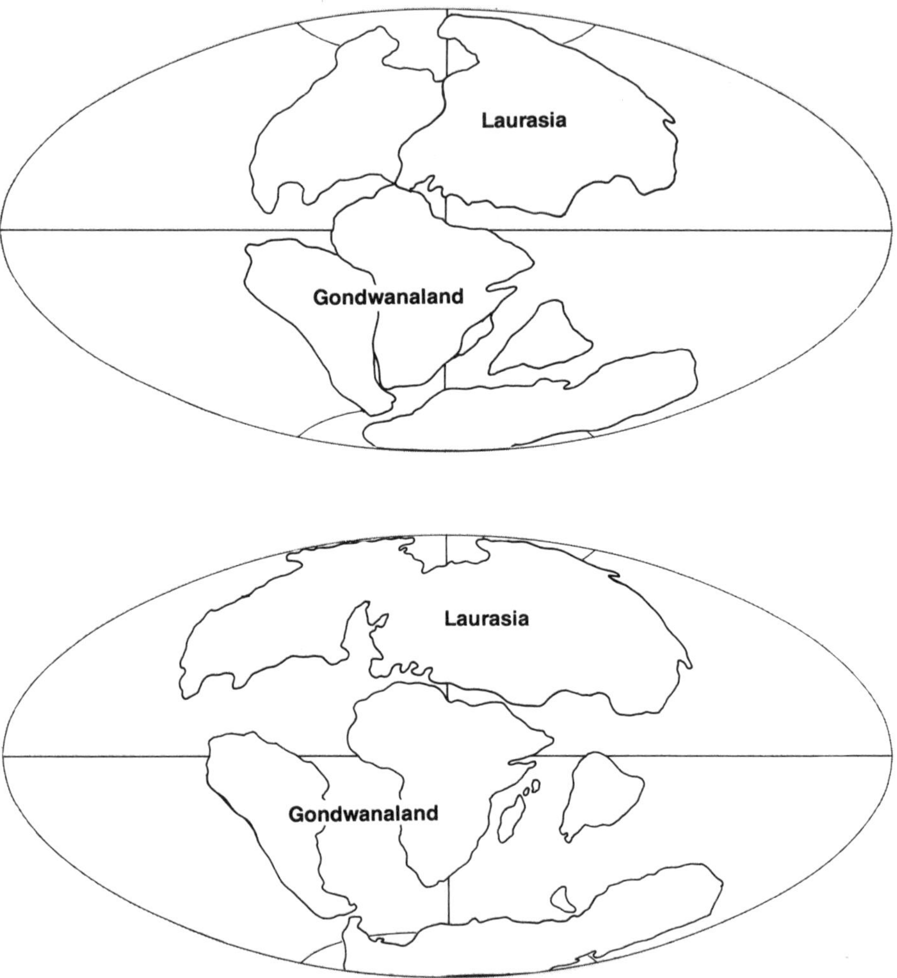

Two stages of the theory of evolutionary continental drift.

Above — the original landmass (Pangaea) is beginning to split into recognisable areas. Laurasia is a combination of Nth America, Europe and Asia. Gondwanaland includes Sth America, Africa, India, Australia and Antarctica. This is estimated at 180 million years ago.

Below — 65 million years ago. A halfway stage to the current shape of the world. Sth and Nth America have yet to join, India is an island and Australia remains attached to Antarctica.

17: *The continental drift theory*

The theory on the continents drifting around the Earth was first propounded by a Frank Taylor in 1908 but it did not take off as an acceptable theory then. It was not until Alfred Wegener who happened upon a scientific paper on the subject, in a library, which advanced this particular theory, that further progress would be made. He was immediately impressed with the idea that continents could drift and from then on he devoted a considerable amount of his time and energy on the subject, lecturing to those who were interested and also to those who were impressed with all the possibilities that such a suggestion raised.

Wegener was lecturing outside his chosen professional field, which was astronomy and meteorology, and because of this he came up against considerable opposition from the geographical societies and scientists in general. For this reason it took a number of years for the idea to be accepted in principle, and before it could be taken to the stage where it became a theory worthy of further serious investigation.

But the change in thinking by geologists and scientists was still very slow in gaining acceptance and it was not until a Harry Hess (who commanded an attack transport during the Second World War which was fitted with a new device called a fathometer that gave a profile of the ocean floor) became interested and it was because of the fathometer that some progress was again made.

He used the device regularly while patrolling the Pacific Ocean and eventually it came to his notice that there were unusual submarine formations projecting upwards from the sea floor with flat tops as though they were old volcanoes that had been eroded at sea level before they slowly submerged beneath the ocean. As a geologist this intrigued him.

To sea floor spreading

However, it was not until after the war that he wrote and presented a paper on his findings and his view that sea floor spreading was the probable cause. From that point onwards research continued at a steady pace with special ships surveying the oceans of the world. By about 1965 it had become an accepted fact that the seabeds of the oceans were spreading more or less steadily from the middle outwards and towards the continents on each side.

At the edge of the continents there are deep trenches where the sea floor is slowly being pushed under the continents. These areas are also the boundaries where the violent earthquakes take place, and it is around the edges of the Pacific especially, where many of the world's volcanoes are situated in the area known as the 'ring of fire'.

The eroded volcanoes discovered by Harry Hess were previously part of an earlier ring of fire before being slowly pushed into deeper water as the sea floor dipped towards the trenches into which they would ultimately disappear. The high mountain ranges on the edge of many continents are the result of this persistent uplifting from the pressure of the sea floor being pushed under the continents (New Zealand's Southern Alps, for instance, are still 'growing').

The reasons that all this information on the continental drift theory is being offered has been to lead up to the established fact that the sea floor of the ocean is now known to spread from near the middle outwards and also that it was the continental drift theory that first promulgated the discovery that the sea floor was actually spreading.

But what has not been established is that the sea floor spreading and the subduction of the sea floor under the continents is equal in that no material is gained or lost. It may have been that the sea floor spreading, although only a matter of an inch or so a year, could still be greater than the amount of subduction taking place.

Over hundreds of millions of years even a gain of say half an inch a year of sea floor spreading over the amount subducted could add up to somewhere near 1287km (800 miles) over a period of about 100 million years. This distance would be added to the circumference of the Earth at the Equator, and of course, the poles. Or if one were attempting to ascertain the circumference of the Earth at some distant period, then of course these kilometres would need to be subtracted. If 1287 kilometres or proportionately less were to be deducted every 100 million years, and this is but a short period of time in the history of this planet, then it would be obvious that this planet could have been much smaller than at present.

With the same amount of atmosphere that is available on Earth now and which should have been available then on a smaller sphere, the greater depth of the atmosphere would be compressed by its own weight and under those circumstances a higher barometric pressure must follow. Also, theoretically, under those conditions the large prehistoric pterosaur would have been able to fly.

Logic suggests that the sea floor spreading must have been, or be, enlarging the size of the Earth. For if the tectonic plates are crashing into other continental plates and forcing the edges of those continents to crumple and form high mountain ranges, which has occurred in the geological past, it indicates that the buckling around the circumference of the Earth has gone in part into mountain building instead of increasing the circumference of the Earth to more than it might have been. This could indicate that the rising internal pressure causing the sea floor spreading was too great, as it was occurring to allow a completely unimpeded increase in the circumference of the Earth horizontally so the buckling of the Earth as mountains was the only other alternative. Without this buckling action through the ages, the Earth of today could have been considerably larger.

18: Greenhouse effect

To round off on the problems being faced in meteorology in the areas of cyclone control, overcoming droughts, and damage by hail, or just attempting to avoid too much rain, a comment or two on the 'greenhouse effect' will not be out of place.

Throughout this book the purpose has been to show a connection between the Moon and weather, and in mentioning the 'greenhouse effect' it is not my intention to prove that the Moon does have an influence in this direction. However, one is inclined to wonder, after reading some of the extravagant statements associated with the so called 'greenhouse effect', whether the scientists have any real understanding on the subject of atmospheric physics. The usual claims on seas rising because of the reduction in ice at the poles, dust bowls for plains from the higher temperatures but never a mention on what happens to the increased precipitation which the extra evaporation must surely bring.

The atmosphere cannot hold all that humidity which must follow the higher temperatures, there has to be a limit to what can be held aloft so it has to fall somewhere as rain. Even if a very high saturation point in the atmosphere may be possible, the result would be the same; moist air over most of the Earth and of course, no dust bowls.

The orbit of the Moon will not be affected by any possible rise in the temperatures so there will be a distinct possibility that the atmosphere will circulate as it does at present and rain will fall as it has done for the past millions of years.

Computers are being used increasingly for this type of climatic study and the projections by them are becoming the authority for making these extravagant claims on the possibility that the world may enter into a 'greenhouse' type climate. But computer experts do say that to believe a computer implicitly is likely to offer false interpretations on whatever subject is under investigation. Caution is necessary on both sides of this subject, is the 'greenhouse' effect factual or will the present status quo prevail. None of us want to make a mistake either way, but any statements bordering on being crazy, we can all do without.

Anyone with an interest in meteorology will have read plenty of material on this subject anyway and they will also know that the 'greenhouse' effect is still but a theory and for the theory to become an established fact, it will take many more years of observations and data collecting.

Unfortunately, no one really knows whether the Earth is warming up or cooling down, and the only way meteorologists, climatologists or scientists will settle the argument that the 'greenhouse effect' does not exist will be for the world to enter into a cooling period. If that occurs, it will settle the discussion once and for all time.

But if the world is going into a warmer period, the problem on whether the warming up is a natural phenomenon or a 'greenhouse effect' will continue to be discussed and it would seem that there can be no finality on which point of view is the correct one.

The theory on the 'greenhouse effect' is based on the increasing amount of carbon dioxide within the atmosphere.

In the late 19th century, the concentration of carbon dioxide in the atmosphere was about 290 parts per million and this figure has increased to something near 330 parts per million at the present time, which is not a high concentration. Oxygen has a concentration close to 25 per cent.

The problem with the increased carbon dioxide levels, apparently, is that the incoming short wave solar radiation will pass through the carbon dioxide but that radiation, after reaching the Earth is then re-radiated outwards as long waves. But they become trapped and will not re-radiate back through the carbon dioxide as efficiently as the short waves had done when entering the atmosphere and continuing on down to the surface. It is said that the trapped long waves cause the atmosphere to become warmer.

The increase in carbon dioxide, it is claimed, is mainly man-made through the burning of fossil fuels such as coal, oil or gas, and this extra quantity of carbon dioxide will only decline through being absorbed by the seas or by plant life generally which takes in the carbon dioxide and converts it into oxygen.

But man has continually cut down the forests which absorb the carbon dioxide, also the seas absorb carbon dioxide only slowly and further, it is stated the use of fossil fuels is on the increase thereby creating more carbon dioxide.

Some predictions do indicate that world temperatures may rise by about 1°C by the 21st century. Others have been stating a rise of between one-four degrees and with the seas likely to increase in height by up to one and a half metres in 50 years time with the flooding of the lower coastal areas.

If these predictions should be anywhere near correct, its effect on the climate will be difficult to assess. Some climatologists have the view that there will be less snow and ice and therefore higher sea levels which will flood low lying coastal districts. That the Arctic and Antarctic ice will melt too, but there seems to be little logic in that statement.

Warmer atmospheric temperatures and warmer seas will cause greater evaporation and more humidity and when this warm saturated air is carried towards the poles, the air will become colder and lose its moisture content as snow in the Arctic and Antarctic regions and it may be possible that the snow and ice will increase in quantity rather than decrease. It must be remembered that both the Arctic and Antarctic are the result of past weather, and not the reverse.

The polar regions will always be frozen because of the six months without sunshine and also because the atmosphere is thinner there, allowing the freezing cold of space to penetrate to ground level.

So before the polar areas can warm up, the space above must also become warmer.

If the 'greenhouse effect' is to become a fact, then the likely effect will be that the equatorial areas will become warmer and also the temperate zones in the parts closer to the tropics, but some of the extra moisture that the atmosphere must contain will certainly be deposited in the area of the frozen north and south poles, and so maintaining or increasing the level of ice there and with the possibility the sea levels will reduce rather than rise.

Another point that seems to be missed is that the higher atmosphere will still remain cold and when the air, super saturated with moisture, is carried up on the high mountain ranges it will deposit that moisture as snow because cold air cannot hold that surplus moisture. Ultimately the snow will compress into ice and glaciers will form or those already there will extend which means that the seas cannot rise in height, and a more likely result will be that the seas will reduce even further in depth.

Climatologists and meteorologists seem to believe that changing temperatures to either colder or warmer will change the present atmospheric movement extensively, when in fact it is the Moon which will dictate how the atmosphere will behave, and while the Moon orbits the Earth as it does now, there will be no alteration in the manner by which the atmosphere will circulate.

As a matter of curiosity from me, and a question that could be answered by some knowledgeable person is — how is it that carbon dioxide, a gas which is the heavier component of air, does not settle in hollows and cause the death of all forms of life in those hollows? Carbon dioxide is an inert gas and does not support any form of life and one might expect that somewhere, in say a valley situation without an outlet, carbon dioxide would settle as a gaseous lake in which there was no life. Wind of course would stir the air occasionally and may reduce the ability for the gas to separate and accumulate in sufficient quantity to become a problem, or that the small proportion of carbon dioxide in the air may not allow the gas to accumulate up to such a height that it would become a danger to some forms of life.

There must be occasions when wind is not in evidence in a valley for many days consecutively, and which could allow carbon dioxide to settle and accumulate for a time at least.

There is another factor which may influence the ability for carbon dioxide to retain sufficient heat which will allow for the commencement of a 'greenhouse effect', and that is the question as to what height will the increased carbon dioxide rise within the atmosphere. It is heavier than air or any of the other gases which make up our atmosphere, so it may be that over the higher country it could be reasonably free of any increase in carbon dioxide, or less proportionally than at the lower levels. Should this be so, then the incoming radiation striking the higher country may still re-radiate outwards as it does at the present time.

The deciding point will be the percentage of higher country to that nearer the sea levels, it may not be that high a percentage but it could still be an escape route for the re-radiation to take.

Although the emphasis is on the inability of the heat rays to re-radiate outwards through the carbon dioxide, the fact is that warm air will rise in any case and it must eventually mingle with the cooler upper air which in turn will cancel out any of the suggested increase in temperatures.

Another factor relates to the changing declination of the Moon over the 18 year cycle. At present (1988/89) the Moon's declination is declining from the maximum of 28 degrees movement north and south of the Equator each month, or five degrees outside the tropics. This extra movement must circulate the warm tropical air further into the temperate zones, thus increasing the average temperatures there. During the next nine years the declination will slowly reduce to 18 degrees north and south of the Equator each month, or five degrees inside the tropics and this will mean that less warm air will circulate north or south of the tropics and so reduce the temperatures in the temperate zones to below the accepted average.

Average temperatures are a split between higher and lower temperatures, so when it is said our temperatures will rise by a stated figure above the average, what will that figure be above the higher temperatures which are used to form the present averages.

We are in a period of naturally higher temperatures because of the present declination of the Moon. As the Moon returns to its minimum declination the temperatures also will decline.

The seas are not increasing in height because the Antarctic ice is not melting from the present higher temperatures; so what of the future? Part of the claimed possible future rise in temperatures, of 2-4°C above the average, must be part of the present higher temperatures, so what is the actual increase these scientists are stating we must expect. If it is but say 1-2°C above the present higher temperatures, then how can the Antarctic ice melt causing the seas to rise by the predicted one or one-and-a-half metres and flood low lying land.

If the maximum declination of the Moon at about 28 degrees in its monthly movement north and south of the Equator is responsible for an increase in the temperatures to above the average, then by researching the weather records for each of the 18 years further back in time, some evidence for a similar rise in temperatures should show in those years.

The latest declination of 28 degrees was in 1988, which means that 1970 or 1971 should give similar high temperatures.

Quoting from the climatological records of the New Zealand Meteorological Service for November: "In some areas this is the 18th successive month with temperatures warmer than normal and 1971 is likely to be one of the warmest years on record for New Zealand."

December was also above the average by 1-1½ degrees and then dropped back to about normal for January 1972.

The next 18 year period goes back to 1952 (with a declination of 28 degrees) or 1953 (a declination of 27 degrees). For 1952 August was above the average, September average to above and October, November and December were all above the average.

August and September of 1953 were above the average, October was average, November was above and December average to above.

The next period back in time with the 28 degree declination was 1932-33. These years are outside the official records that are held by me, so it was necessary to obtain whatever information that was published in the *Waikato Times* of Hamilton, New Zealand, on any weather phenomena then.

Commencing with November 1931 it was said of that month the temperatures were the highest for a November for a number of years. December was the third successive dry month in North Otago with three inches of rain in that period. It was also very dry in Nelson and along east coasts generally.

In South Australia the heat was extreme for four days with temperatures at 100°F and with a further comment that temperatures were 110-115°F in New South Wales, the hottest for 51 years, which is close to 3 x 18 years.

For January 1932 there was a reference to droughts in Hawke's Bay, dry in Canterbury and North Otago and rain urgently needed in Waikato. In Brisbane, Australia, a heat wave was reported there.

In February a comment said the pastures in Waikato were almost useless. In Hawke's Bay the reference was to blazing heat.

In the summer of 1932-33 for November, Auckland was dry with 26.9mm (1.6") rain recorded and 40.4mm (1.59") for Cambridge. The comment in Hamilton for December said the weather was sultry with rain required urgently and with the added remark that it was the driest for 18 years. That year, 1914, recorded the lowest rainfall total for Hamilton ever 730.25mm (28.75"). The yearly average is about 1219mm (48").

In January 1933 both Marlborough and Nelson were dry. Dunedin had a heat wave with temperatures up to 99°F and Waikato also had a warm week.

February in Australia; a heat wave in Perth with temperatures of 100°F over four days and a further comment that the heat wave was the most prolonged in history. Sydney sweltered day and night for a time with temperatures at 110°F.

The earliest official records held by myself were for the year 1934 when the Moon's declination was about 26 degrees plus. For the month of December it was said some districts had temperatures by as much as 7°F above the average.

January 1935 was the hottest January ever experienced and February was also much above normal with continuously high temperatures. March was considerably above the average with one exception — Hawke's Bay.

April was the sixth consecutive month with above average temperatures which meant that November 1934 was also above the average. May was stated to be below the average.

Could it be said that all these years were part of an earlier 'greenhouse effect'?

The earliest comment I have seen relating to the 'greenhouse effect' was an article in the *NZ Listener* in 1976 and only an odd reference otherwise until the 1980s. Since then the supposition that the 'greenhouse effect' could be factual has snowballed to the stage that it has become an obsession in the scientific and meteorological circles.

In the early 1970s meteorologists and climatologists were speaking about the coming ice age.

Methane

Since this chapter was written, Dr Dave Lowe of New Zealand, in August 1986, stated that methane — a flammable, colourless, odourless gas — was on the increase in our atmosphere. He said the increase inexplicably commenced to build up concentrations in the Elizabethan age, slowly at first, but gathering momentum from about 1900 onwards.

Methane comes from natural gas reservoirs underground, from decaying plants and animals.

The percentage in the atmosphere is small but associated with carbon dioxide, it is stated the methane assists in trapping the Sun's rays.

Dr Lowe said that in the next couple of centuries the ice caps at the poles will melt, the climate will become hot and uncomfortable for life, and Earth would be a very, very hostile place on which to live.

It seems illogical for methane to build up concentrations in the 1600s only. Why not earlier? Also, how can one be certain that the proportion of methane in the atmosphere was not higher at some other age in the Earth's history.

We always seem to have this threat of melting ice with every new discovery or investigation into a build up of whatever gas there is in our atmosphere.

Methane would be the newest and strangest phenomenon to add to the list. It is flammable and surely it will be burnt out of the atmosphere before large concentrations could cause any of the problems we are supposed to expect.

In the cooler regions on Earth, fires are alight for heating in many homes. Some factories use coal or oil fired boilers. Steel mills use intense heat, so why the worry about a methane build-up?

Lightning, possibly, would create more flash for burning methane than all man-made fires on Earth.

This seems to be a story for news value rather than being factually a problem within our atmosphere.

19: *Weather forecasters and forecasting*

Every country in the world has its departments of meteorology where daily weather forecasts are compiled and issued to the media for public consumption. In some instances the forecasters are unknown, hidden away from the public gaze, so that if a forecast should be incorrect, he or she as an unidentifiable person, can blush in the privacy of their workplace and the public can only shake their fist in the air in frustration.

Very seldom will a forecaster apologise for a bad forecast, he probably shrugs his shoulders and keeps his embarrassment to himself.

In some countries the forecasters become front persons on television and explain the weather forecast in fascinating detail by pointing to the various areas likely to be affected by rain, snow, thunderstorms, heat or cloud; in other words the weather of tomorrow.

If the forecast is near correct on the day following, then on that night, when giving the next forecast, reference can be made brashly "As I said last night", "As expected" before continuing on with the forecast. But if the previous forecast should be incorrect, nothing is said, no excuse, no apology, no reason for the failed forecast will be offered.

The public, through taxation, pay for a weather forecast. They pay the high salaries the forecasters receive, but the public cannot admonish them.

Many letters to the editors of newspapers show how frustrated the public can be when a forecast for, say, fine weather for a long holiday weekend is soured by wet weather.

In this modern technological age the meteorological services have the very latest aids to assist in giving the most up to date forecast it is possible to give. Forecasts are upgraded during the day, and one may ask, why is it necessary? A forecast given in the evening, either on radio, television, or published in the newspaper should be spot on for the next day, but often enough it is out. Rain may be forecast to commence the next morning and it arrives the following day. Or the weather is to clear through the day and it is still raining the day after.

Before the advent of satellite pictures an excuse could be made, even although an incorrect forecast may have inconvenienced the public.

Today it should not occur.

The satellite pictures are available regularly through the day. Observers in many parts of a country ring in, or in some instances radio their observations into the central meteorological office with barometric pressures, temperatures, the type of cloud and visibility and also if it is wet or fine. There are also trained meteorologists in other main centres to pass on the local conditions.

Yet in spite of all of this assistance, the forecasts can be wrong. Actually it is the observers or recorders around the country who tell the forecasters what the weather is like. The forecasters only put it together — without the recorders the forecast would probably be wrong more often than not. The early morning weather forecast given before the recorders send in their observations are often incorrect for this reason. The forecasters do not know that a change in the weather has taken place.

This would not be an isolated example

There was one particularly bad example which occurred in New Zealand not long ago from the point when this episode is being mentioned and when every modern aid was available to produce an accurate weather forecast.

The weather was fine and mainly sunny from an anticyclone over the country. The forecast given on radio at 7.15am for that day was short with little variation for all of New Zealand, and the outlook for the following day was equally short with the same fine conditions continuing.

The radio announcer, referring to the forecaster by name, said something like: "That was a nice short forecast. I wish we could have more of them." The forecaster in his wisdom replied: "I like to do the best I can for the farmers and their haymaking." It was that time of the year when haymaking was in full swing.

The next day dawned cloudy with rain, it rained all day and into the evening. Around the countryside there were hectares upon hectares of cut hay lying in saturated paddocks. It was a very expensive forecast for the farmers. The forecaster's name and the exact date that this event occurred has been filed away.

What is a forecaster's responsibility?

Every now and again a yacht will founder or be dismasted. Many yachties have complained that the weather forecast they received did not, in any way, match the conditions encountered.

How many mountaineers get trapped by bad weather when the expectation was for the weather to be sufficiently fine for them to be able to complete the climb.

Forecasting for snow to low levels is another area which is not covered adequately. The snow actually has to be falling before mention of its arrival will be made in the forecast. This is especially a problem for sheep farmers at lambing time. Thousands of lambs will be lost because a farmer does not receive sufficient, or any warning, that snow is on the way.

At other times sheep may be on higher slopes and they will be trapped there by a heavy overnight snowfall when some warning would have given the farmer time to bring the sheep down to paddocks closer to haystocks and better shelter. Night snow is particularly dangerous because snow falls silently and by the time a farmer knows it is snowing it may be too deep to shift stock easily.

Flooding is another area not adequately handled. Insufficient time or no time at all will be given which will enable a farmer to shift stock to higher ground.

A farmer may not like the sound of the rain he hears, but unless some warning is offered on the magnitude of the rain in the catchment areas, he may not do anything until it is too late, and also unfortunately very heavy flood rains seem to occur overnight when it is difficult to do anything anyway.

Meteorological services seem to adopt the policy that if they give statistical information after the event it will cover up for their inadequacies before the event.

Information on where snow fell and how deep, what stock losses there were, on the floods, how deep they were, how much rain fell and whether it was a more serious flood than a previous one. Everything will be publicised afterwards but absolutely nothing or little beforehand.

Sometimes as a knee jerk action, and for a while afterwards, events which could possibly lead to a reoccurrence of a flood, snow storm or whatever, will be preceded by a warning on the possibility for a repeat of that type of situation which was not forecasted for on the previous occasion, but this warning will slowly die away until another non-forecasted flood or heavy snow storm arrives.

Because meteorology is considered a science which the public does not fully understand, it also follows that a Member of Parliament who becomes the Minister in Charge of Meteorology will not have a grasp of the subject either, so he cannot contradict a director in charge of a meteorological organisation who may unnecessarily ask for extra funds, or request extra equipment. The Minister cannot ask penetrating questions on why so many are employed in producing so little.

A meteorological section becomes an ivory tower and an impenetrable castle designed to confuse anyone who attempts to understand its working.

The hierarchy are suitably titled as director, assistant director, a director of this section or that section, until it reaches the stage that no one is a plain employee. For each such ranking a few more thousand dollars in the salary is the reward.

What a weather forecast really means

If it is necessary to know what the weather was like in any area away from one's own locality during the previous day, listen to the first weather forecast for the new day, because what you will hear will be just that, yesterday's weather handed out as today's weather forecast.

The subsequent forecasts throughout the day will catch up on the changing situation when new information arrives, so one will still receive as a forecast the weather as it is at any place at that particular time.

If one is in doubt on this point, just listen to one's local forecast change as the weather changes about you. Often, should there be more than one radio station available to listen to, one can receive contradictory weather forecasts at any one time and this will be confusing to the listener.

For some reason it is called a forecast when in actual fact it should be called a postcast of the previous weather or perhaps better still the service should be called 'The Weather Information Service' leaving out the word forecast.

The accuracy of the official weather forecasts every now and again comes up for scrutiny and as late as 1983 British forecasters came in for a drenching for being 42 per cent correct.

This is an amazing admission when their weather office has all conceivable modern equipment from satellite coverage to high powered computers and probably, like New Zealand, an army of recorders throughout the country who actually tell the weather office what the weather is like at the time the recorders make a report.

Have the forecasters lost the art of physically doing a synoptic weather chart and now rely entirely on a computer printout?

Everyone knows that a computer does not think and it will only process accurately any accurate information fed into it. If the operator does not have the correct information on hand, then the results will be less than useful as a weather forecast.

As one British meteorologist once admitted, a farmer, probably, will have a better idea on the weather situation in his own locality than any meteorologist located miles away in his office.

In New Zealand a spokesperson for the Meteorological Service stated in 1984 that 25 per cent of their predictions were wrong, this in spite of employing at least 50 weather forecasters.

Fifty weather forecasters or more for a population of just over three million people, take out the children and those people who have very little interest in the weather, that is a high rate of forecasters to the remainder of the interested population and just for a 75 per cent accuracy rate.

Average rainfall figures

Most or possibly all meteorological services reduce rainfall statistics to averages, with both monthly and yearly figures. As it has been mentioned the phases of the Moon do play an important part in our weather, along with other factors. Most months will have only one each of a particular moon phase in a month but every now and again a phase at the beginning of the month will be repeated again at the end of the month making five moon phases in that month.

The repeated moon phase can be any one of the phases, new, first, full, or last. Occasionally a similar type of weather situation will repeat itself from the same moon phase at the beginning of a month and again at the end of the month. It does not matter whether the situation is low pressure and unsettled or anticyclonic and fine but it does mean a doubling up of that particular type of weather in one month.

Then that month is included in the statistical information to create an average for all those months. If a month with the same two moon phases in it happens to be very unsettled it is then compared with the average rainfall and entered as being X number of mm above the average. This will be statistically correct but it is not really an accurate comparison. Of course the opposite situation will arise if a month is dry, it will be below the average by whatever mm the average rainfall is for that month.

Months with only one each of the moon phase per month will commence the month with any one of the phases and this situation will determine which part of the month may be dry or wet.

One may ask, of what use are averages unless it is for statistical value only for from those statistics one cannot say what type of weather will be produced in any one month should an attempt be made to project those statistics into the future.

A similar situation to that produced by the moon phases will occur with the apogee and perigee cycle. These cycles follow a similar pattern to the moon phases where apogee will be at the beginning of a month and occasionally about once in a year, again at the end of the month (which is the same for perigee). And once again, whatever the weather may be at the commencement of the month could be repeated at the end of the month be it unsettled or fine weather. With the Moon at perigee and when the weather is unsettled under those conditions, then when perigee does occur twice in the month, the rainfall can be excessively heavy, so much so that records for that month could be broken. Statistics will record that fact but not the reason why it was made possible for the record to be broken.

20: University research

In 1976 a group of professors, scientists and students from the University of Waikato in New Zealand, claimed push button weather will soon be a reality, all based on the study of tree rings, and claiming that past climatic history from the tree rings, the ratio of carbon, hydrogen, and oxygen isotopes vary with different temperatures, and using the carbon 14 dating method, they will determine what the weather was like during a particular season in the past. The information fed into computers will, by some imaginative way, project into the future with details on what the climate will be like during some stated period.

It was a good article but something seems to be missing. How will the computer transfer the details of, say, 50 years ago and project them to whatever year sometime hence? If this was possible, then there would be no need to look at tree rings. The actual weather on a day from the past could be fed into a suitably programmed computer and the computer would be able to state exactly what day in the future would have that same weather. The computer would have to be programmed to do the projection whether for tree rings or by using daily weather details, so if the computer had been so programmed, the method for the projections must be known also, and the projection would work equally well for tree rings or from past daily weather records.

It is now almost into the 1990s and no more has been heard on this research. What has it cost the taxpayer for this futile effort?

This is only one example where money is wasted on some project or other where the individual puts forward a case for research in an area and on a subject only the intelligent think will be of some use in the future.

However, every university in the world will sponsor similar researches, but this is not to suggest that no good ever comes out of research within the university environment.

There was the story which came out of Britain in 1977 on 10 years research on why an egg shell split when the egg was boiled.

Then there was the grant by the British Government for two years research into the design of toilet seats. Impossible? It happened.

However, it is recommended that should a meteorologist have an ambition to become the director in a research department of meteorology, or even the director of a meteorological organisation in the near future, he would be well advised to study astronomy, especially where it relates to the Moon's orbit and all aspects of the variations within that orbit.

21: *Astronomers — have they failed us?*

I feel that the true reason for the lack of progress in really understanding the movement of the atmosphere lies at the door of the observatories or the people within.

Astronomers could be accused of having their heads in the clouds. They frequently make weather-related comments without making an attempt to solve the weather problem.

The earlier comments in this book show that if astronomers had used their specialised knowledge sensibly, the movement of the atmosphere would have been closer to being understood very much earlier in the history of meteorology, and meteorology would have been very much further down the path to a more complete weather knowledge than it is at present.

Astronomers studying sunspot activity were able to relate the sunspot cycle to Jupiter's orbit, but could not find a reason why these spots developed or the reason for Jupiter being in a particular position on its orbit when they appeared.

Astronomers in 1977 found evidence of water in the so-called spiral nebula M-33, 2.2 million light years away, but they still cannot tell us about the water within visible sight in our own atmosphere, or how it behaves.

In 1975 it was mentioned that the Moon was moving away from the Earth in what was referred to as the 'constant of gravity'. It was concluded that the reduction of gravity could result in the planets, Sun and the Moon increasing in size. I have already spoken of this in a previous chapter when commenting on the extinction of dinosaurs, etc.

Someone from the Victoria University in Wellington, New Zealand, mentioned that the Moon's orbit was increasing by one cm each year. He added that there was little likelihood that we would lose the Moon, but that he thought that we could live without it. He did not know much, did he?

The gravitational attraction of the Moon to the Earth was known, but never related to its possible effect on the atmosphere. If it had been taken into account, meteorology may have advanced faster than it has done over the past 100 years.

The effect the Moon has on the spin of the Earth, on its axis, by friction and so slowly increasing the length of the day was known, but nothing on the effect of friction on the Earth's atmosphere. If this possibility had been mentioned, meteorology may have incorporated the suggestion, and who knows, a better understanding on the formation of cyclones may have been the outcome.

Sometimes it may be said that familiarity breeds contempt, and this thought may well have been applied to the Moon; it was there, it has been studied and there was a great big universe out there full of mystery. The atmosphere was a nuisance because it distorted, or refracted, that which was being viewed through it and only hindered the astronomers in their work.

It was a pity that astronomers made positive statements then called them facts on matters unknown, and on points non-provable on the activity around the universe, when a closer look near at hand would have produced meaningful results.

It may not have pleased meteorologists to have had astronomers or scientists interfering in their work, for it is a case of each discipline to its own but in this instance, when meteorological discoveries were in their infancy, astronomers may have been more successful as meteorologists had they applied their minds intelligently to the subject. Astronomers had the required knowledge, meteorologists did not.

22: *Quotes and sayings on weather interpreted*

Weather has always universally been a topic for conversation. In the days before a better understanding of the weather and its possible causes, the more observant in our midst used sayings based on observations to explain weather changes. These observations related to such things as cloud formations, the look of the sky or animal behaviour, or the way trees blossomed or developed during a season.

Spring time was a particularly important season. Any signal, of whatever kind, was grasped in an attempt to foretell the likely conditions for the coming summer season.

The population of every country would interpret the signs in a way that was relevant to their particular country.

Fishermen and sailors would scan the sky for signs that were important to them, such as those foretelling the possible arrival of a storm.

The appearance of the new moon was greeted with particular interest, especially as the Moon began to change from new and to begin to show the first signs of its crescent.

If the weather had been dry, the angle of the crescent could indicate the possibility that water would spill over from its cusp and be the beginning of a period with more rain.

A newspaper columnist asked an astronomer if the position incorporating the first appearance of a new moon in the evening sky could be predicted, that is whether upright and thus symbolically emptying water, or on its back with both its cusps pointing upwards and thus holding water, in which position it was suggested dry weather would follow. The astronomer said it was not possible to predict how the Moon would come in.

This intrigued me sufficiently to follow this thought through. Each month the Moon's position was noted as the crescent first appeared in the sky, showing the new moon phase. After about three years, it became obvious that the position of the crescent was very much the same for each lunar month for approximately the same time of the year (allowing for the fact that the new moon does not fall on the same date of a month in each year).

So it would seem unlikely that the position of the new crescent could, in itself, offer any indication on what the weather would be for the ensuing days or month. If this were so then, whenever a new moon came in on its back, the weather would be dry, and this dryness would reoccur at about the same time each year. This, we know, is not correct.

Meteorologists, in general, are inclined to scoff at any suggestion that the Moon influences the weather. The Moon *does* influence the weather, but not because of its shape at any given time.

One quote on the Moon and the weather goes this way:

> *The moon and the weather*
> *May change together,*
> *But a change of moon*
> *Does not change the weather.*

New Zealand, in common with other countries, has its own peculiar quotes and sayings. Other sayings having a universal appeal have been imported.

One belief is that when the cabbage trees bloom profusely, it will be a long dry summer. The cabbage tree has a long trunk with a corklike bark and a head of long narrow leaves similar to flax. In the crown of the leaves can be obtained a soft core which it was thought the Maori cooked as we would do with cabbage, but it is understood this belief is not correct.

Another tree native to New Zealand, the pohutukawa (also referred to as the Christmas tree, because it usually blooms about then), has a bright red blossom with a petal structure similar to a thistle. The tree is hardy and grows along the shoreline in the north of the North Island. In some years the blossom is scrappy, and in other years the trees are covered in a mass of blossom. This sign is also said to herald a good summer, i.e. not too wet.

Another omen is the early flowering of flax.

No one has yet stated that *all* these blossom indications need to occur together before a supposed drier spell, or if only one tree or plant is sufficient to indicate the probability. What are the interpretations supposed to mean when some of the dry summer indications don't agree, one with the other at any one time? No one has yet offered an opinion on what to expect. Perhaps it means a mixed season.

However, to use these possibilities as a firm guide for the coming season is to bring disappointment to oneself at some stage. History will show that such predictions are not based on fact. A probable reason for profuse blooming is that the trees and plants had a favourable autumn or winter.

In some districts, the residents are favoured with a particular shaped hill or mountain. One conical in shape offers the best effect where cloud will give an indication on what the weather may be in the short term.. If the crown of the hill can be seen with the cloud higher and above it, the weather will be fine, but as the cloud descends on to the mountain or hill, the locals will expect rain to develop. The lower the cloud descends the more certain it will be that rain is close.

The reason is that the cloud will act as a barometer, remaining at a buoyancy level (or the level which may be at the boundary of a temperature change). As the atmospheric tide descends, or becomes lower, the buoyancy level of the cloud will drop lower, until the mountain or hill is covered with cloud.

A common saying on this condition is: "If you can see the mountain there is going to be rain, if you cannot it is raining."

Every now and again the sky has a very clear appearance. Binoculars seem unnecessary or they are an added bonus. Sometimes this condition occurs after rain and sometimes before. If this phenomenon is after rain, a common comment will be that the sky has been washed clean, but this explanation does not explain the phenomenon occurring before rain. An atmospheric tide can produce the same effect. When the atmospheric tide is low, the air molecules are less compact, which allows one to see further. If this condition is before rain, it can be an indication that rain is near, possibly a day away.

When the night sky is without a moon and clear of cloud the stars appear very close. This is another indication of a low atmospheric tide: if there is no moon up in the sky, it must be on the other, or daylight, side of the Earth while at the new moon phase; or it is still to rise, as it will be when it is in the last quarter rising near midnight and also when in the first quarter setting near midnight.

A top meteorologist in England was reported as saying that, despite sophisticated aids such as computers, satellites or radar, a farmer was better advised to study the sky each day for himself and make his own assessment on likely weather.

'If the sun rises red and fiery — wind and rain' can be translated as 'red sky in the morning shepherd's warning, red sky at night shepherd's delight'.

'Rain before seven, fine by eleven' has a practical answer, for the rain will come in on an atmospheric low tide during a first quarter moon which has yet to rise, and the atmospheric tide will rise higher by 11am with fine or clearing weather. In reverse, of course, one often sees a cloudless sky before 7am but rain by 11am. The moon phase will be in the last quarter which will be setting by about 11am which will bring in the rain if it is around.

Fruits and berries are sometimes mentioned as a basis for future seasonal weather predictions. When the skins are tough it is suggested a hard winter is ahead. The answer may be that the season just past was too dry!

Many people, in particular those whose livelihood is from the land, have noted signs which inform them of the possible changes in the weather. Unfortunately many of these signs are not advertised or passed around for general use. The observations have been kept within districts or within the family.

If as many as possible of these observations could be made available for universal usage, we would all benefit.

One comment received by me stated that dew on the grass indicated fine weather and with no dew, rain could be closer.

Washing slow to dry on the line on a windless day may also be an indicator that rain is close; a humid atmosphere finds it difficult to absorb any more moisture from the wet clothes.

Many people suffer aching joints a few days before rain arrives, and once the rain commences, the joints cease to ache. This may relate to a highly electrical or magnetically charged atmosphere affecting people. It is a custom for many of those who suffer arthritis or rheumatism to wear a copper bracelet, which does not quite form a circle, to ease the problem. Rain often arrives during either of the quarters of the Moon and, as mentioned in another chapter, the gravitational attraction of the Moon and Sun concentrated in a 90 degree segment of the sky could affect the Van Allen Belt (a magnetic force circling the Earth) by bringing it closer to Earth and so increasing the magnetic charge in the atmosphere which in turn may increase the magnetism within a person's body to a level they cannot tolerate.

When kingfishers, small birds normally found near water, are in the area, the saying is: 'The nearer the house, the nearer the rain'.

Bird fanciers say that their birds become restless 24 hours before rain.

Sheep will stay under trees, while cows may stop eating.

Cats also seem to know when rain is close.

Ants, it is claimed, run in circles when rain is about, but move in straight lines for fine weather.

A Polish sea captain who kept hamsters, said that when they were quiet and lazy a storm was coming but, if they were active, the weather would be fine.

It is said that migratory birds flying from one hemisphere to another will not leave until the weather systems are suitable, with wind directions to assist them on their flight. How do they receive the advance knowledge necessary to fly those vast distances? Do they have a better understanding of what makes the atmosphere tick than we do? If migratory birds do have this understanding, then the time is long overdue for humans to catch up with them and gain this knowledge.

There are one or two rhymes from the days of sail which still have a place on land, for they indicate when stronger winds can be expected. One referring to a rising barometer is:

> *When rise begins after low,*
> *Squalls expect and clear blow.*

This rhyme would refer to the leading edge of an anticyclone advancing after a depression or cold front has passed by and the barometer will begin to rise.

Another rhyme goes this way:

> *Hens scratching and mares' tails*
> *Make tall ships carry small sails.*

This type of cloud, cirro stratus, is usually seen in the earlier mornings and in time to anticipate that stronger winds may be likely that day and it is also likely to be a time when there is no moon up in the sky, indicating a low atmospheric tide.

Some country folk suggest one should observe the direction in which a hedgehog builds its nest — "The front to the north, or south, east or west, for if 'tis true what common people say, the wind will blow the quite contrary way."

Another saying is: "A cow with its tail to the west makes weather the best, a cow with its tail to the east makes weather the least." This could be a quote from the Northern Hemisphere.

From most of what has been written on weather signs and folk sayings, one cannot help but wonder if we are so lacking in confidence and knowledge on matters meteorological that we are prepared to give meteorological intelligence to trees, birds and animals. We are prepared, under normal circumstances, to admit that birds and animals generally act by instinct, but seldom by intelligence, so why do we, on this one subject, give them intelligence?

23: Conclusion

In writing this book, it was never intended to comment on the usual methods for forecasting the weather in detail other than where it was in contrast with the theories expounded. For the gathering of information from around the country (such as the temperature readings, barometric pressures, humidity, the type of cloud, its cover and probable height, the wind direction and strength) is necessary before a sea level weather map can be compiled.

The physical movement of the atmosphere can be ascertained at the time it is being reported, and the past movements have been recorded as weather maps, together with the written weather forecast for that period. It is assumed that these forecasts are corrected or notes attached giving the actual weather at that time.

After reading this book, some may feel that they should have done so in conjunction with one of the 'orthodox' books which are available so as to gain an understanding on the Moon theory as it affects our weather. However, those who have read it to this point in most instances will be well conversant with 'orthodox' meteorology.

There are any number of books available offering explanations on how depressions circulate and how the associated fronts behave, also an explanation on what type of weather may be expected when conditions are anticyclonic.

The behaviour of the winds and why they flow from the direction that they do will be explained.

The books can be varied in the type of information or subject matter offered. Some will specialise in climatic history, others will offer detailed information on weather forecasting, the terms used and what they all mean.

There are a few books which will cover every aspect on weather from past ice ages and probable causes, the distribution of weather, how the atmosphere circulates and what causes it to circulate as it does, the way hail is formed and snow, including descriptions of the many beautiful patterns that snow flakes will form into, and the temperature range necessary to form the various patterns.

Also, the many droughts which affect parts of the world, when they occurred and the misery they cause, the seasons of the year will be explained and what the equinoxes mean, the development of meteorology and its history may be included and also an explanation on the efforts to make numerical forecasting more viable.

There are plenty of books offering the 'orthodox' view but most, if not all of them, from now on will not be of any further use as text books on meteorology.

The world ought now to be entering a new and knowledgeable meteorological era.

Harry Alcock became interested in long range weather forecasting after commencing an umbrella manufacturing and retail business in 1958.

Umbrellas were in demand only when it rained, and a problem surfaced when advertising them for sale. The local newspaper required copy for an advertisement several days before it appeared in the newspaper. It was the custom to advertise in the Monday's evening paper which gave the remainder of the week to gain a result. Copy for the advertisement was required to be in the office on the Friday, so it was five days later before the advertisement could produce a result.

If it remained dry after the advertisement appeared in the newspaper, the cost of the advertisement showed a loss. If it rained, the advertisement showed a profit. So it seemed logical to try and eliminate the loss.

The original object was to attempt to anticipate the weather for about one week ahead which took in the week following the appearance of the advertisement.

If there was to be any success in discovering what the weather may be in the future, the method most likely to be successful could be based on that which gives us sea tides and which are motivated by the Moon.

It was reasoned that if the sea had tides, then the atmosphere must have tides also. However, all the meteorological literature said there was no such thing as an atmospheric tide, or if any, it was minimal.

The next step was to obtain books on the Moon and its orbit. A study of the books indicated that there could be a link with the Moon and the weather.

From then on it was trial and error, which has continued until fairly recent times.

After gaining a reasonable accuracy rate, the forecasts were developed in a manner where they could be acceptable and sold to farmers and other interested parties.

A number of weekly publications in rural districts receive and use the weather forecasts and also an odd monthly paper.

Because weather is so important to everyone world-wide, it was decided that this book should be written in an attempt to initiate further research on the Moon's effect on the atmosphere and weather, for atmospheric tides are weather.